STANDARDIZATION IN MEASUREMENT: PHILOSOPHICAL, HISTORICAL AND SOCIOLOGICAL ISSUES

History and Philosophy of Technoscience

Series Editor: Alfred Nordmann

Titles in this Series

1 Error and Uncertainty in Scientific Practice
Marcel Boumans, Giora Hon and Arthur C. Petersen (eds)

2 Experiments in Practice
Astrid Schwarz

3 Philosophy, Computing and Information Science
Ruth Hagengruber and Uwe Riss (eds)

4 Spaceship Earth in the Environmental Age, 1960–1990
Sabine Höhler

5 The Future of Scientific Practice: 'Bio-Techno-Logos'
Marta Bertolaso (ed.)

6 Scientists' Expertise as Perfomance: Between State and Society, 1860–1960
Joris Vandendriessche, Evert Peeters and Kaat Wils (eds)

Forthcoming Titles

The Mysterious Science of the Sea, 1775–1943
Natascha Adamowsky

Research Objects in their Technical Setting
*Alfred Nordmann, Astrid Schwarz, Sacha Loeve and
Bernadette Bensaude-Vincent*

Reasoning in Measurement
Nicola Mößner and Alfred Nordmann (eds)

STANDARDIZATION IN MEASUREMENT: PHILOSOPHICAL, HISTORICAL AND SOCIOLOGICAL ISSUES

EDITED BY

Oliver Schlaudt and Lara Huber

Routledge
Taylor & Francis Group

LONDON AND NEW YORK

First published 2015 by Pickering & Chatto (Publishers) Limited

2 Park Square, Milton Park, Abingdon, Oxfordshire OX14 4RN
52 Vanderbilt Avenue, New York, NY 10017

Routledge is an imprint of the Taylor & Francis Group, an informa business

First issued in paperback 2020

BRITISH LIBRARY CATALOGUING IN PUBLICATION DATA

Standardization in measurement: philosophical, historical and sociological
issues. – (History and philosophy of technoscience)
1. Measurement – Standards.
I. Series II. Schlaudt, Oliver, editor. III. Huber, Lara, editor.
530.8'1-dc23

ISBN-13: 978-1-8489-3571-6 (hbk)
ISBN-13: 978-0-367-59876-1 (pbk)

Typeset by Pickering & Chatto (Publishers) Limited

CONTENTS

List of Contributors   vii
List of Figures and Tables   xiii

Introduction – *Lara Huber and Oliver Schlaudt*   1
Part I: Making the Field
  1 'A Branch of Human Natural History': Wittgenstein's Reflections
    on Metrology – *Martin Kusch*   11
  2 Metrology and Varieties of Making in Synthetic Biology
    – *Pablo Schyfter*   25
  3 Refrain from Standards? French, Cavemen and Computers: A (Short)
    Story of Multidimensional Analysis in French Prehistoric
    Archaeology – *Sébastien Plutniak*   39
Part II: Standardizing and Representing
  4 The Double Interpretation of the Equations of Physics and the Quest
    for Common Meanings – *Nadine de Courtenay*   53
  5 An Overview of the Current Status of Measurement Science: From the
    Standpoint of the *International Vocabulary of Metrology* (*VIM*)
    – *Luca Mari*   69
  6 Can We Dispense with the Notion of 'True Value' in Metrology?
    – *Fabien Grégis*   81
Part III: Calibration – Accessing Precision from Within
  7 Calibration in Scientific Practices which Explore Poorly Understood
    Phenomena or Invent New Instruments – *Léna Soler*   95
  8 Time Standards from Acoustic to Radio: The First Electronic Clock
    – *Shaul Katzir*   111
  9 Calibrating the Universe: The Beginning and End of the Hubble Wars
    – *Genco Guralp*   125
Part IV: The Apparatus of Commensurability
  10 Measuring Animal Performance: A Sociological Analysis of the Social
     Construction of Measurement – *François Hochereau*   139

11 The Measure of Democracy: Coding in Political Science
   – *Sharon Crasnow* 149
12 Measuring Metabolism – *Elizabeth Neswald* 161
Part V: Standards and Power – The Question of Authority
13 The Social Construction of Units of Measurement: Institutionalization, Legitimation and Maintenance in Metrology – *Hector Vera* 173
14 A Matter of Size Does Not Matter: Material and Institutional
   Agencies in Nanotechnology Standardization – *Sharon Ku and*
   Frederick Klaessig 189
15 Measuring by which Standard? How Plurality Challenges the Ideal
   of Epistemic Singularity – *Lara Huber* 207

Notes 217
Index 251

LIST OF CONTRIBUTORS

Nadine de Courtenay is Associate Professor at the Université Paris Diderot in the Department of History and Philosophy of Science and member of the laboratory SPHERE, Paris, France. She has a PhD in solid state physics as well as in philosophy of science. She has taught atomic physics, solid state physics and measurement science for many years as an Associate Professor at the Conservatoire National des Arts et Métiers. Her research interests are centred on the history and philosophy of nineteenth- and twentieth-century physics. She focuses more particularly on statistical mechanics, Ludwig Boltzmann's philosophy of science, scientific modelling and the philosophy of measurement.

Sharon Crasnow is Professor of Philosophy at Norco College in Southern California. Her primary research interests are epistemological questions surrounding methodology, particularly the use of case studies, mixed methods research and feminist philosophy of science. Recent publications include 'The Role of Case Study Research in Political Science: Evidence for Causal Claims', *Philosophy of Science*, 79 (2012), pp. 655–66, 'Feminist Philosophy of Science: Values and Objectivity', *Philosophy Compass*, 8 (2013), pp. 413–23, and 'Feminist Perspectives', in N. Cartwright and E. Montuschi (eds), *Philosophy of Social Science: New Directions* (Oxford: Oxford University Press, 2014). She is also the co-editor of *Out from the Shadows: Analytical Feminist Contributions to Traditional Philosophy* with Anita M. Superson (Oxford: Oxford University Press, 2012) and *Philosophical Feminism and Popular Culture* with Joanne Waugh (Lanham: Lexington Press 2013).

Fabien Grégis is a PhD student and teaching assistant in the history and philosophy of science at Université Denis Diderot (Paris 7). He works on the notion of 'measurement uncertainty' in experimental sciences, with an emphasis on contemporary metrology and physics. The main epistemological issues he addresses are about the nature and function of physical quantities, the role and interpretation of probabilities in measurement science, the experimental testing of theories and the articulation of measurement with scientific realism, in connection with a study of the recent history of metrology.

Genco Guralp is a PhD candidate in philosophy at Johns Hopkins University. His research interests include historical epistemology, philosophy of science and science and technology studies. His dissertation examines the discovery of the acceleration of the universe by two independent research teams, using this as a test case to probe the debate between the Bayesian and the error-statistical philosophies of science. He has also written on the philosophical implications of the recent developments in empirical cosmology: 'Cosmology and the End of Weberian Science', in B.-J. Krings, A. Schleisiek and H. R. Zabaleta (eds), *Scientific Knowledge and the Transgression of Boundaries* (Dordrecht: Springer, forthcoming 2015).

François Hochereau is a research fellow in sociology at INRA (the French National Institute for Agricultural Research). His research interrogates the role of measurement as a 'social operator' through support of knowledge, assessment scale and rule of action. This work has led him to focus on the construction, consolidation and legitimation of new references of knowledge or assessment in agriculture, including the evaluation of crop standards in wheat breeding, the establishment of a docility test in bovine breeding, the social construction of animal pain in livestock systems and how new genomic techniques alter measurement configurations within plant and animal selection procedures.

Lara Huber is a philosopher and historian by training. Her areas of specialization are philosophy and history of science and technology, philosophy of the life sciences, phenomenology and image science/aesthetics. Currently, her research focuses on standardization and its impact on concepts of scientific knowledge. Recent publications within this realm are 'Mutant Mice. Experimental Organisms as Materialised Models in Biomedicine', *Studies in History and Philosophy of Biological and Biomedical Sciences*, 44:3 (2013), pp. 385–91 (with L. K. Keuck) and 'Colors of the Brain. Investigating 14 Years of Display Practice in Functional Imaging', *Neuroimage*, 73 (2013), pp. 30–9 (with M. Christen et al.).

Shaul Katzir is a senior lecturer at the Cohn Institute for the history and philosophy of science and ideas, Tel Aviv University. He has held a number of research fellowships in Europe and Israel. His research focuses on the history of physics and technology and their interactions in the nineteenth and twentieth centuries, on which he has published extensively. His article here is part of a larger project: 'From Sonar to Quartz Clock: Technology and Physics in War, Academy and Industry' which explores the applications of piezoelectricity and related technologies in World War I and its aftermath, including the development of time and frequency standards.

Fred Klaessig is currently with Pennsylvania Bio Nano Systems, a small firm focusing on reference materials used in investigating chromatographic effects at the nanoscale. Earlier, he was the Technical Director and also Business Director for the Aerosil Business Line of Evonik Degussa GmbH. His assignments ranged from commercial overview (Product Management, Production, Sales) to technical responsibilities involving customer support, new product introduction, liaison with the R&D Department in Germany and regulatory matters. AEROSIL is a trade name for fumed silica, which has been manufactured for sixty years and which is often cited as an example of a nanomaterial. Fumed silica, fumed titania and other fumed metal oxides are utilized in many fields for reinforcement, rheology control, abrasion and UV absorption. In recent years, the great interest in nanotechnology has raised safety and registration concerns about materials of this class. These issues, both everyday technology and EHS, led to his involvement in ASTM (E56), ISO (TC229) and industry organizations addressing these broader topics. Fred received a BSc in chemistry from the University of California, Berkeley and a PhD in physical chemistry from Rensselaer Polytechnic Institute. His earlier industrial experiences were with Bio Rad Laboratories as a Quality Control Chemist and various R&D management positions at Betz Laboratories, now a division of GE Water Services, where his responsibilities involved scale, corrosion and microbiological control in many chemical industrial processes.

Sharon (Tsai-hsuan) Ku is an Assistant Research Professor at the Department of History and Politics, Drexel University, and a guest research fellow at the Office of History, National Institutes of Health, USA. Her research focuses on the material culture in science and science policymaking. This approach is driven by her early career as an experimental physicist, and her current involvement in two international standard organizations ASTM E56 and ANSI/ISO TC229 as a STS scholar. She is particularly interested in real-world practices of precision and standardization, and the social and ethical values embedded in standard making. She has developed several ethnographic studies in National Institutes of Standards and Technology, National Cancer Institute, and two NSF-funded Centers for Nanotechnology in Society, investigating the socio-technical infrastructure for technology standardization, translation and interdisciplinary collaboration between science and social science. Her current project examines the knowledge infrastructure for data-intensive nanotechnology, particularly the social epistemology of database construction for nanorisk framing and quantification. Ku received a BSc and MSc in physics from National Chiao Tung University, Taiwan, and a PhD in history and philosophy of science from the University of Cambridge in 2009.

Martin Kusch is Professor in Applied Philosophy of Science and Epistemology at the University of Vienna. His main areas of interest are social epistemology, philosophy of the social sciences, sociology of science, history of psychology and Wittgenstein's philosophy. Recent books include *Knowledge by Agreement: The*

Programme of Communitarian Epistemology (Oxford: Clarendon Press, 2002) and *A Sceptical Guide to Meaning and Rules: Defending Kripke's Wittgenstein* (Chesham: Acumen, 2006).

Luca Mari (MS in physics, 1987; PhD in measurement science, 1994) is a Full Professor of measurement science with Università Carlo Cattaneo (LIUC), Italy. He is the author or coauthor of several scientific papers published in international journals and presented at international conference proceedings. His research interests include measurement science and system theory. Dr Mari is currently the chairman of the Technical Committee 1 – Terminology of the International Electrotechnical Commission (IEC) and an IEC expert in the Working Group 2 (International Vocabulary of Metrology) of the Joint Committee for Guides in Metrology (JCGM).

Elizabeth Neswald is Associate Professor for the history of science and technology at Brock University, Ontario. She has written on the cultural history of thermodynamics (*Thermodynamik als kultureller Kampfplatz. Zur Faszinationsgeschichte der Entropie, 1850–1915* (Freiburg i. Brsg.: Rombach, 2006)), the media philosopher Vilém Flusser (*Medien-Theologie. Das Werk Vilém Flussers* (Köln: Böhlau-Verlag, 1998)), and science in nineteenth-century provincial Ireland, as well as publishing numerous essays on the early history of nutrition physiology and the work of Francis Gano Benedict. She is currently completing a monograph on *Counting Calories: Thermodynamics, Statistics and the Making of Modern Nutrition Science*.

Sébastien Plutniak was formerly trained both as an archaeologist and a social scientist and is a PhD candidate in sociology of science at the EHESS (School for Advanced Studies in the Social Sciences), Toulouse, France. His main interests include the dynamics of science activities, the formalization and mathematization of scientific fields and the relations between disciplines.

Oliver Schlaudt is Associate Professor at the philosophy department, University of Heidelberg, Germany, and member of the Poincaré Archive, Nancy, France. His research focuses on epistemology. He recently completed a book on pragmatist theories of truth: *Was ist empirische Wahrheit?* (Frankfurt a. M.: Klostermann, 2014).

Pablo Schyfter is a researcher in science and technology studies. His first work in the field concerned gender and technology. Using the sociology of knowledge, feminist theory and poststructuralism, he developed a framework for understanding the ontological constitution of gendered subjectivities, sexed bodies and technological artefacts. More recently, he has studied the ongoing consolidation of synthetic biology. He has written about the field's understanding of design, its epistemic practices and the ontology of synthetic biological artefacts.

He intends to continue studying the field and the particularities of engineering metrology. Presently, Pablo is a Lecturer in science, technology and innovation studies at the University of Edinburgh.

Léna Soler is Associate Professor at the Université de Lorraine and a member of the Laboratoire d'Histoire des Sciences et de Philosophie – Archives Henri Poincaré, in Nancy, France. Her areas of specialization are philosophy of science and philosophy of physics. She studied physics first (with an Engineering degree in materials science from the Formation d'Ingénieurs de l'Université de Paris Sud Orsay) and went on to study general philosophy and philosophy of science at Université de Paris I–Sorbonne, France. Her PhD, completed in 1997, was on Einstein's early work on light quanta and theorizing processes in physics. In 2008, she received a four-year research grant from the French Agence Nationale de la Recherche for the project 'Rethinking Sciences from the Standpoint of Scientific Practices'. She wrote an *Introduction à l'épistémologie* (Paris: Ellipses, first edition in 2000; revised and enlarged edition in 2009), and has been the main editor of several volumes: *Rethinking Scientific Change and Theory Comparison: Stabilities, Ruptures, Incommensurabilities?* (Dordrecht: Springer, 2008), co-edited with H. Sankey and P. Hoyningen; *Characterizing the Robustness of Science: After the Practice Turn in Philosophy of Science* (Dordrecht: Springer, 2012), co-edited with E. Trizio, T. Nickles and W. Wimsatt; *Science after the Practice Turn in the Philosophy, History, and Social Studies of Science* (New York: Routledge, 2014), co-edited with S. Zwart, M. Lynch and V. Israel-Jost; a special issue on contingency (*Studies in History and Philosophy of Science*, 39 (2008), co-edited with H. Sankey); and a special issue 'Tacit and Explicit Knowledge: Harry Collins's Framework', *Philosophia Scientiæ*, 17:3 (2013), co-edited with S. D. Zwart and R. Catinaud.

Hector Vera holds a PhD in sociology and historical studies from the New School for Social Research. He is a researcher at Instituto de Investigaciones Sobre la Universidad y la Educación, Universidad Nacional Autónoma de México. His doctoral dissertation, 'The Social Life of Measures: Metrication in Mexico and the United States, 1789–1994', is a historical-comparative analysis on how diverse institutions and groups (state agencies, scientific societies, chambers of commerce and industry) appropriated and signified the decimal metric system. He is the author of *A peso el kilo. Historia del sistema métrico decimal en México* (Mexico: Libros del Escarabajo, 2007), a monograph on the adoption of the metric system in Mexico. He is also the author of 'Decimal Time: Misadventures of a Revolutionary Idea, 1793–2008', *KronoScope: Journal for the Study of Time*, 9 (2009), pp. 29–48, and co-editor, with V. García-Acosta, of a volume on the history of systems of measurement, *Metros, leguas y mecates. Historia de los sistemas de medición en México* (Mexico: CIESAS, 2011).

LIST OF FIGURES AND TABLES

Figure 3.1: Citations of methodological references and global transitivity   49
Figure 3.2: A model for the development of prehistoric archaeology   51
Figure 8.1: The multivibrator   114
Figure 8.2: A triode-maintained tuning fork   115
Figure 14.1: Illustration of the procedure of NIST SRM production   195
Figure 14.2a: NIST gold RM   198
Figure 14.2b: Report of investigation   199

Table 3.1: Trends of research in French prehistoric archaeology   47
Table 6.1: A suggestion for a classification of quantities   90
Table 7.1: The instrument frame applied to the equal arm balance   98
Table 7.2: The simple exemplar of calibration in UNSI practices   100
Table 7.3: Calibration in PUNSI practices   105
Table 14.1: Size as a multi-value parameter in the case of gold RM   197

INTRODUCTION

Lara Huber and Oliver Schlaudt

This volume of fifteen essays offers the reader a multidisciplinary approach to standardization in measurement.[1] Measurement is crucial to modern civilization, and standardization is crucial to measurement. Standardization in measurement is also a challenge, for it is a multidimensional object in the study of which the epistemic and the social are intertwined and, in last analysis, cannot be separated. Over the last few decades research in the natural and life sciences has been marked by an unstable and often tense relationship between philosophical, historical and sociological approaches. The days of overt aggression (e.g., the 'science wars') seem to be over, but even if historians, sociologists and philosophers are interested in collaborating with one another, they usually do not know how, so there is a general tendency to withdraw into their own traditional domains. We think that standardization in measurement offers a quite natural opportunity to overcome disciplinary boundaries. This volume seeks to inform the reader about the fundamental relationship between measurement and standardization and to explore standardization in measurement in its various aspects: standardization of procedures, instruments and objects, of units of measurement and of vocabulary.

Measurement and Standardization

Measurement

Measurement is an old companion of mankind, dating back at least to ancient Mesopotamia. It is also intimately linked to science. Some have even argued that the development of quantitative methods is coextensive with science itself. '[In] any special doctrine of nature', Kant tells us in his *Metaphysical Foundations of Natural Science* from 1786, 'there can be only as much *proper* science as there is mathematics therein' – and thus measurement, if the mathematics is to be linked to experience.[2] Viewed realistically, however, measurement is surely neither sufficient nor necessary for science; rather, it is key to widely differing scientific practices. As such it is strongly oriented towards the development of precision instruments and statistical analysis as much as data mining techniques. Not surprisingly, measurement attracted much attention in the early days of epistemology when physics was still the queen of the sciences – it is extremely prominent in Mach, Poincaré, the early Carnap and the Vienna Circle – and only ceased to do so when measurement theory, at intervals from the 1890s on but predominantly since the 1940s, tended to be treated in a purely formal way (axiomatic theories of measurement, theory of scales). The formal approach to measurement – conceived as the representation of objects by numbers – studied the construction of mappings between a given 'empirical relational structure' and a numerical counterpart, neglecting thereby the intricate role of laboratory work involved in accessing the 'empirical relational structure' in such a way that numbers can be mapped onto it. This work is often of a local nature, not stabilized once and for all, and rests on material artefacts inherited from tradition and adapted to novel use. These contingent, 'history-laden' circumstances of measurement are mirrored in the full expression of a measurement result, consisting not only in a numeral – as supposed by the formal or 'representationalist' approach – but also in a unit and a margin of error. These aspects begin to attract attention from various perspectives: studies of error, historical epistemology, the practical turn. In this way, measurement re-enters the scene. Indeed, a recent review article observes a return of measurement to the forefront of philosophical research:

> A wave of scholarship has emerged in the past decade that views measurement from a novel perspective, bringing standards, artefacts, modelling practices, and the history of science and technology to bear on philosophical problems concerning measurement. This recent work departs from the foundationalist and axiomatic approaches that characterized the philosophy of measurement during much of the 20th century. Inspired by developments in the philosophy of scientific modelling and experimentation, contemporary authors draw attention to scientific methodology and especially to metrology, the science of measurement and standardization.[3]

Standardization

Crucial to this aspect of measurement is standardization. Standardization is a practice of regulation that extends into all spheres of human action. Standardization in nineteenth- and early twentieth-century industrial production can be seen as a major event in human cultural and social history. Accordingly, standardization cannot be reduced to exclusively scientific purposes such as measurement and its implications for related practices in everyday life. There are very different objects of regulation and topoi of standardization, including the stabilization of material objects as much as the control of human interactions. It is for this reason that sociologist Lawrence Busch in his recent book speaks of standards as 'recipes for reality'.[4] Still, it is the domain of science where the genuine nature of standards is most apparent, given that they are both prerequisites for scientific practices and outcomes of scientific expertise. Standards formalize and regulate strategies of validation and therefore contribute significantly to the evolution of scientific practices as such. Standards, as for example clinical practice guidelines, provide trust in scientific methodology but also prioritize a given set of research practices.[5] Several studies in the history of science and technology have shown that practices of standardization arise predominantly on the basis of interactions with technical devices, notably measuring instruments. In his book *History of the Thermometer and its Uses in Meteorology* historian W. E. Knowles Middleton at least implicitly illustrates that any history of measurement simultaneously gives insight into a history of standardization.[6] This characteristic is not restricted to manufacturing technology or the calibration of measuring devices but includes quite varied approaches to formalization in science. It also responds to significant regulatory challenges, as explored, for example, by Geoffrey C. Bowker and Susan Leigh Star in their social study on the classification of disease.[7] Actually, it is the social sciences that have shown a steady and substantial interest in practices of regulation and the effects of standards on human action – or rather on human self-perception in the course of practices of normalization.[8] There is a huge literature on how standardization impacts on how individuals are viewed and judged, including, most famously, *Discipline & Punish: The Birth of the Prison* by Michel Foucault and *The Mismeasure of Man* by Stephen Jay Gould. As regards practices of standardization in laboratory science, the concept of 'standardized packages' introduced by sociologist Joan H. Fujimura could serve as a starting point for further research into the extent to which the standardization of technical devices serves a stabilization of facts and is effectively a means of regulating human action across divergent areas of application.[9]

Up until recently, the systematic challenges posed by standardization have not been addressed from the perspective of philosophy of science. Allan Franklin's book *Experiments in Particle Physics in the Twentieth Century* could be read as a systematic case study of how standards of measurement (here standard deviation) determine epistemic values such as significance or credibility. Due to technical innovation, standards remain objects of improvement or even displacement. This aspect could be classified as part of the history of scientific progress. Additionally, the case of 'shifting standards' could also be framed as a problem associated with an established scientific practice. As a consequence, epistemic values, such as significance or credibility, may be affected when a given standard is challenged due to its modified use.[10]

Standardization in Measurement: A Multidisciplinary Approach

Standardization in measurement is thus a crucial and yet largely neglected component in the production of scientific knowledge. Furthermore, it is a multidimensional issue, which lays bare the entanglement of the scientific, technological and social issues that come into play in the development of knowledge in the natural and life sciences. As such, its study offers an excellent opportunity to rethink disciplinary boundaries. What we hope to gain from this new perspective on standardization in measurement is a much clearer picture of how scientific, technological and social issues not only coexist but indeed interrelate with and mutually influence one another in science.

Multidisciplinary approaches run counter to the commonplace distinction between facts and (epistemic) values, between a contingent context of discovery and a self-sufficient context of justification. This distinction is constitutive of epistemology as understood by many philosophers, and it is behind the endless quarrels between philosophers and sociologists which have taken place over the last forty years – and which, we think, have often enough paralysed efforts to understand how science works. The fact/value distinction has always had its doubters too, however. The 'heretical' movement of pragmatism, for example, challenged the fact/value dichotomy in a number of fundamental ways, while some early representatives of the sociology of science were also sceptical about it. Karl Mannheim, in his 1929 work *Utopia and Ideology*, regarded the fact/value dichotomy simply as an over-hasty, hypostatizing institutional strategy aimed at establishing epistemology as an independent discipline.[11]

This critical stance has reappeared more recently. In the wake of Nelson Goodman, Catherine Elgin proposed in 1989 'the Relativity of Fact and the Objectivity of Value', i.e., the thesis that fact and value 'are inextricably intertwined'.[12] Feminist philosopher Lynn Hankinson-Nelson and, in neo-pragmatist mode, Hilary Putnam both draw parallels between their attack on the fact/value distinction and Quine's critique of the synthetic/analytic dichotomy: the normative and the descriptive, just as the analytic and the synthetic, might well be aspects of our epistemic engagement with the world, but they are not mutually exclusive categories to which all the individual items of our knowledge can finally be allocated. The reason for this is that statements do not express bare facts but rather entangle facts and conventions – and so they do also with facts and values.[13] In her book *The Fate of Knowledge*[14] (2002) Helen Longino seeks to overcome the dichotomy between the (non-social) rational and the (non-rational) social by identifying the underlying values constitutive of scientific discourse and thus of the production of scientific knowledge. Far from being merely a disturbing factor (Francis Bacon's famous 'idols of the mind'), here the social *is* the rational – not only in the trivial sense of an instantiation of the rational in the social, but in the sense that the social is constitutive of the rational, or at least of one of its constituents.

The case of standardization in measurement is a rather striking instantiation of this observation. To take a case in point, fixing a unit of measurement is clearly a social act, and it also has a social purpose, namely, to record a finding and to enable its communication and independent reproduction. It has often been stressed that ratios of (homogeneous) quantities are independent from units and hence do not contain conventional elements. True, but this is not the end of the story because, viewed realistically, scientific knowledge cannot be reduced to mere ratios and their relations (cf. Nadine de Courtenay's paper for further details on what follows here). If, for example, we write down a law of nature that links quantities of *different* kinds as a classical proportion, and if we wish to apply mathematical operations to it, then we have to fix the denominators of the ratios so as to transform the classical proportion into an *equation* which holds between the numbers. Thus conventionally fixed units come into play again. And if we further wish to give an invariant expression to this equation, or, to put it another way, if we want exactly the same equation to hold between both the magnitudes and their measures (Maxwell's 'double interpretation' of equations as it is tacitly assumed in today's science), even a 'coherent set of units' has to be constructed – coherent in itself, but also coherent in the context of contemporary physics and its experimental means. Accordingly, understanding scientific knowledge as it actually occurs within scientific practice demands the study of the 'invisible infrastructure' of metrology;[15] more generally, it requires a recognition of the contingent, 'history-laden' context in which scientific knowledge is embedded and with which it is entangled in various intricate, often improvised ways. Objectivity, to put it in general terms, cannot be accounted for as a general trait of scientific knowledge without taking into account the local and provisional strategies adopted in order to achieve it. It emerges in the superposition of different practices – recording, communicating, reproducing – and should in the last instance not be regarded as an independent property over and above them.

The case of standardization in measurement hence takes us even further than Longino did in her study, as she focuses only on theory formation in the sense of a choice between empirically equivalent theories relating to a given set of data. What we are concerned with, in measurement and standardization, is the very production of data itself. Standardization in measurement seems to provide an excellent opportunity for demonstrating the interrelatedness of epistemological and social factors and thus the need for an approach that combines philosophical, sociological and historical studies and leaves behind the unfruitful quarrels which have characterized debates in the past.

Dimensions of Standardization in Measurement

This volume gathers together fifteen contributions by metrologists, sociologists, historians and philosophers. In order to enable the entanglement of the social, the historical and the epistemic, or of the practical and the theoretical, to 'show itself' (as Wittgenstein might say), we have deliberately *not* arranged the papers according to their disciplinary aims or origin but have rather grouped them according to their shared themes (or at least those themes highlighted most prominently in them). As a general guide to this volume, we have identified various sets of questions in relation to standardization in measurement: Who are the *actors* of standardization? How do they secure acceptance of their norms and how do they maintain the authority to do so – what are the mechanisms involved? What is the *subject* of standardization: artefacts such as units, prototypes or measurement devices, procedures and methods (i.e., protocols), or language, symbolic representation and vocabulary, both on an empirical and on a methodological level (in metrology)? What are the *effects* of standardization? Standardization contributes towards stabilizing data and validating knowledge, to creating uniform artefacts and shaping scientific practices. Furthermore, standardization can impact on community building by addressing disciplinary boundaries. Here both descriptive and normative aspects clearly appear, though we hold them to be present also in the preceding questions. Standardization is seen to play a crucial role in both constructing fields and disciplinary identities and in justifying knowledge. Of course, none of the contributions to the present volume covers only one of these aspects. The five themes according to which they are grouped are as follows:

Making the Field

Measurement is a social practice based on instruments and technical expertise. In this respect, then, standardization in measurement could be understood as a coordinated and approved exercise of rule following in a socially and technologically shaped space. Hence concepts and values of measurement affect social practices: Marin Kusch opens the volume with a philosophical paper on Wittgenstein's rudimentary 'sociology of metrological knowledge' implicit in his use of metrological metaphors in the analysis of grammar. The relationship between measurement and the social sphere, however, is a dialectical one, given that the establishment of shared standards in measurement – of units, artefacts and methods – is also a community-building practice. With regard to scientific communities, it may help in constructing or enhancing disciplinary identity – or in reflecting on existing disciplinary boundaries, as Pablo Schyfter's paper illustrates. Schyfter analyses how synthetic biologists seek to promote the discipline of biological engineering by identifying existing 'standardized parts' in genetics and how this practice is linked to producing valid engineering knowledge. Whereas standardization could be regarded as essential for community building, we might nevertheless question the extent to which the introduction of standard formats impacts on the goals of research in a given field. Sébastien Plutniak's sociological study, for example, analyses the role of 'multivariate analysis' of data in the disciplinary dynamics of French prehistoric archaeology as well as its impact on the communities' definition of its epistemic object.

Standardizing and Representing

Standardization in measurement, though not generally recognized as such in epistemology, is crucial to representing nature by means of mathematics, numbers and equations (formalization I). However, it also demands conceptual efforts in metrology, and thus the science of standardization in measurement standardizes its own vocabulary as well (formalization II). In her philosophical paper Nadine de Courtenay reveals the metrological preconditions for the representation of physical relations in mathematical equations and shows how this metrological basis of physical theory links to the coordination of scientific, technical, economic and everyday activities. Luca Mari reflects on the current status of measurement science, taking as a basis the development of the *International Vocabulary of Metrology* (*VIM*) from 1984 to the present day. The existence and the development of the *VIM* reflect practical demands as well as a changing conception of measurement. Fabien Grégis takes up a current issue of the *VIM* debate when he asks: 'Can we dispense with the notion of "true value" in metrology'? He relates the rejection of the notion of 'true value' in the most recent *VIM* to an epistemic turn in metrology, characterized by a focus on practical issues. He lays bare the entanglement of metaphysical and epistemological issues within measurement science.

Calibration: Accessing Precision from Within

Devices, apparatuses, techniques and procedures have to pass calibration tests before entering the stage. These tests often exhibit a recursive structure: they depend on both the past and the future, the theoretical and the practical, on science and metrology. Léna Soler provides the outlines of a theory of calibration. She shows that even in the everyday use of well-mastered instruments, calibration can be a delicate procedure, and she analyses the challenges involved in elaborating new practices and prototypes. In his historical study, Shaul Katzir shows how new practices of standardization impact on the understanding of scientific values. He traces the history of frequency standards from the 1920s to the invention of the quartz oscillator. Here we witness how the material prerequisites of a popularization of precision came into being in the twentieth century.

Genco Guralp's case study on the beginning and the end of the 'Hubble Wars' explores the rivalry between two alternative calibration schemes in the quest to determine the Hubble constant, and shows how the conflict became meaningless when in the 'precision era' a new material culture developed, aimed at reducing error rather than demanding commitment to a single method. He shows how this error reduction programme displays a structure of epistemic iteration.

The Apparatus of Commensurability

Quantification, coding, evaluation – how can scientific phenomena be accessed by measurement and to what extent does the establishment of standards reflect possible answers to this question? François Hochereau shows in his case study on measuring animal performance in breeding how establishing a metric, understood as a collective process, makes things governable and at the same time operates as a device for reaching agreement, how it provides information but also brackets out aspects of the objects it is applied to. Sharon Crasnow addresses a case of measurement in the political sciences: in her analysis of coding as a way of measuring latent concepts she shows that quantification can be goal dependent, i.e., pragmatic decisions within quantification may work for one goal but may not be easily transported to different contexts. Convergence in the sense of Hasok Chang's epistemic iteration thus could not be guaranteed for cases such as measuring democracy. Commensurability as an epistemic end of standardization might be challenged if the area of application of a new standard is expanded: Elizabeth Neswald, coining the notion 'apparatus of commensurability', pinpoints measurement in nutrition physiology as a historical chapter in the intersection of the physiological and the social sciences where different dimensions of measurement strengthen but also undermine one another.

Standards and Power: The Question of Authority

Standardization sets rules, i.e., technical and scientific norms. It thus demands sovereignty over actants, human or non-human, artefacts and their users. Standardization presupposes social and political authorities. Hector Vera, following Peter L. Berger and Thomas Luckmann, investigates how units of measurement are socially constructed, arguing that they are a bundle of different processes such as institutionalization, sedimentation and legitimation, all of which are at work in socially and disciplinary heterogeneous groups. Standards express authority and are discussed as powerful but ambiguous non-human actors: Sharon Ku and Frederick Klaessig challenge the ideology of exact measurement in nanotechnology by analysing the micro-structure of standardization: the so-called Gold Nanoparticle standard is at once a 'harmonizing' calibration device and an 'irritating object' at the research frontier. The relationship between standards and power is a multi-layered issue: Lara Huber explores the purposes for which standards are introduced and how standard formats are established. This also includes the question of how the very status of standards – their 'epistemic singularity' – might be challenged in parallel regimes.

Taken together, these fifteen papers offer plentiful evidence of why standardization in measurement is a central issue. Even as they address a wide range of perspectives on standardization, including procedures, units of measurement and basic vocabulary (such as metrological nomenclature), these fifteen papers are held together by one common topic: although the epistemic and social status of standards might be scientifically approved, it is by no means beyond question. This may be so for purely technical reasons, but it may also be due to internal scientific debates arising from the need for effective standardization as a means of accessing phenomena through measurement.

1 'A BRANCH OF HUMAN NATURAL HISTORY': WITTGENSTEIN'S REFLECTIONS ON METROLOGY

Martin Kusch

Introduction

In this paper I want to defend two theses. According to the first, Ludwig Wittgenstein's occasional remarks on metrology give content and support to the idea of a 'sociology of metrological knowledge'. According to the second thesis, Wittgenstein's remarks on metrology, relativism and rule following give support to Bas van Fraassen's 'constructive-empiricist form of structuralism'.[1]

My paper has four sections. I shall begin by showing that metaphors of measuring are a pervasive feature of Wittgenstein's work, and that they allow us to identify key elements of his thinking about metrology. Subsequently I shall argue that the way in which Wittgenstein leans on Einstein's clock coordination as a metaphor for rule following supports: (i) a 'communitarian' and 'finitist' reading of rule following; and (ii) an analysis of measuring in sociological rule following terms. It is these two ideas that first and foremost bring Wittgenstein into the proximity of a 'sociology of metrological knowledge'. I strengthen this link by proposing that Wittgenstein's thinking about metrology is best summed up as a form of 'metrological relativism'. Finally I shall turn to van Fraassen. I shall propose that he is right to lean on Wittgenstein in defending a subject- or agent-centred philosophy of measurement.

Analogy and Beyond

In this section I shall briefly review the central metrological analogies in Wittgenstein's work from the 1930s to the 1950s. I leave aside the interesting metrological ideas in the *Tractatus* since explaining the latter would take too much stage setting.

I begin with the area of grammar and language. Here the most important analogy is that between *unit of measurement* and *rule of grammar* on the one hand, and *result of measurement* and *empirical proposition* on the other hand:

> The rules of grammar are arbitrary in the same sense as the choice of a unit of measurement.[2]

> The relation between grammar and language is similar to that between deciding upon the metre as the unit of length and carrying out a measurement...[3]

One idea Wittgenstein seeks to make plausible with this comparison is that rules of grammar are *arbitrary* – or at least: *as arbitrary as, and not more arbitrary than,* choices of units of measurement. Later in this paper we shall see that this condition limits the arbitrariness of rules of grammar considerably.

Another metrological analogy important in Wittgenstein's reflections on grammar is that between language and ruler: 'To express something in the same language' means to use the same ruler'.[4] In order to compare different measurements of length we need to know the ruler or scale upon which the measurements are based. Analogously, in order to understand individual sentences we need to be able to situate them in the context of a language to which they belong. This thought gestures towards a form of 'meaning holism'.

Turning from grammar to mathematics, two key passages are the following:

> Rules of deduction are analogous to the fixing of a unit of length ... [Wittgenstein] thought ... that the comparison ... would ... make you see that [rules of deduction] are really neither true nor false.[5]

> Geometry and arithmetic ... [are] comparable to the rule which lays down the unit of length. Their relation to reality is that certain facts make certain geometries and arithmetics practical.[6]

One implication of the analogy is that, like units of measurement, mathematical propositions or rules cannot be evaluated as true or false in any correspondence sense. Instead they must be thought of as being more or less useful within the institution of mathematics. A related point is put forward in terms of an intriguing analogy between logical or mathematical proofs and the role of the standard metre in Paris:

> If I were to see the standard metre in Paris, but were not acquainted with the institution of measuring and its connexion with the standard metre – could I say, that I was acquainted with the concept of the standard metre? Is a proof not also part of an institution in this way?[7]

In other words, to understand the standard metre is to understand its (former) role in our metrological institutions. Likewise, to understand a proof is to understand its function in an area of mathematical practice.

Sometimes Wittgenstein uses 'archive' (i.e., the location of the standard metre) to refer metaphorically to a 'social location' or a 'social status'. This is the social status of being in the common ground of mathematicians: 'A calculation could always be laid down in the archive of measurements. It can be regarded as a picture of an experiment ... It is now the paradigm with which we compare...'.[8] That is to say, a calculation ceases to be an experiment and becomes a paradigm when it is given the social status of being indisputable.

The status that accrues to calculations when they are 'deposited' in the common ground of mathematics also influences what we take these calculations to be about. In depositing them in the archive we stop treating them as being about worldly objects and start thinking of them as being about *nothing but numbers*. Decisions about the reference of mathematical terms thus follows decisions about their social status:

'20 apples + 30 apples = 50 apples' may not be a proposition about apples. ... It may be a proposition of arithmetic – and in this case we could call it a proposition about numbers. ... When it is put in the archives in Paris, it is about numbers.[9]

As far as Wittgenstein's account of 'certainties' ('hinges' or 'hinge propositions') is concerned, I have argued elsewhere that the general distinction 'unit vs. result of measurement' is central here, too.[10] Certainties are like units of measurement, and empirical propositions are like results of measurements – accordingly certainties cannot be true or false in a straightforward correspondence-theoretical sense.

There also seem to be other respects in which Wittgenstein's thinking about metrological units and standards informs his reflections about certainties. Indeed, there seems to be a parallel between his famous claim that the standard metre cannot properly be said to be, or not to be, one metre long, and his proposal that certainties are best not regarded as things we can know or doubt. Certainties function as standards of reasonableness, and such standards are not self-predicating.[11]

A further central theme of *On Certainty* is that agreement in certainties is essential for shared knowledge. Wittgenstein often illuminates this theme by referring to the importance of clock coordination for the determination of simultaneity across locations. The destruction of certainties 'would amount to the annihilation of all ... yardsticks';[12] and: 'Here once more there is needed a step like the one taken in the theory of relativity'.[13]

The analogy between clock coordination and agreement in responses is not an altogether new theme in *On Certainty*. It played a significant role in Wittgenstein's comments on colour classification and calculation already in the early 1940s:

The certainty with which I call this colour 'red' is the rigidity of my ruler. ... My investigation is similar to the theory of relativity since it is, as it were, a reflection on the clocks with which we compare events.[14]

> The clocks have to agree: only then can we use them for the activity that we call 'measuring time'. ... One could call calculations 'clocks without time'.[15]

Up to this point I have focused on metrological phenomena as analogues or models of, or metaphors for, other phenomena. It remains for me to document the further idea that the relationship between metrological and other phenomena is more than just an analogy. Two respects stand out. First, the fixing of units of measurement usually happens through grammatical rules. As we shall see, this link allows us to throw new light on measurement from the perspective of the rule following considerations:

> A rule fixes the unit of measurement; an empirical proposition tells us the length of an object (And here you can see how logical analogies work: the fixing of a unit of measurement really is a grammatical rule, and reporting a length in this unit of measurement is a proposition that uses the rule.).[16]

The second and related way in which the relationship between metrological and other phenomena is more than an analogy for Wittgenstein is that samples – standards, prototypes – are thought of by him as parts of language. They are grammatical items and thus part and parcel of what makes a shared language possible: 'It is most natural, and causes least confusion, to reckon the samples among the instruments of the language'.[17]

To sum up, for the later Wittgenstein (proven) mathematical propositions, rules of grammar and certainties can be understood on the model of units or standards of measurement. Accordingly, they are not true or false in a correspondence-sense, and best thought of as more or less practical. Wittgenstein uses the metrological analysis to sketch a sociological account of mathematical, linguistic and epistemic phenomena: to be a unit or standard of measurement is to have a social status within an institution. In the next two sections I shall follow this sociological theme more closely.

Einstein and the Rule Following Considerations

Einstein was important for Wittgenstein in more than one way. In this section I am interested primarily in Wittgenstein's use of Einstein in the context of the rule following considerations. The following lines are central:

> ... Someone asks me: What is the colour of this flower? I answer: 'red'. – Are you absolutely sure? ... The certainty with which I call this colour 'red' is the rigidity of my ruler... When I give descriptions, that is not to be brought into doubt...
>
> Following according to the rule is *fundamental* to our language-game. It characterizes what we call description.
>
> My investigation is similar to the theory of relativity since it is, as it were, a reflection on the clocks with which we compare events.[18]

In an important paper Carlo Penco interprets this passage in the following way. Einstein's coordinate systems stand to invariant laws as Wittgenstein's cultural systems stand to rule following: 'as we use physical invariants and systems of transformation for comparing different coordinate systems, we may find in the human ability of rule following the universal medium through which we may compare different cultural systems'.[19] While I have benefitted from Penco's analysis, I am inclined to put the emphasis differently. I submit that in the quote given Wittgenstein stresses the general and silent agreement that makes rule following possible. This aspect is of course central to Saul Kripke's well-known interpretation of the rule following considerations: 'If there was no general agreement in the community responses, the game of attributing concepts to individuals ... could not exist'.[20] The importance Wittgenstein gives to 'the common behaviour of humankind' when analysing the possibility of rule following shows that Kripke is on the right track.[21] These commonalities are the backdrop against which different language games can be 'tabulated' – or 'measured' – in Wittgenstein's 'übersichtlichen Darstellungen' ('well-ordered synopses').[22]

Moreover, the analogy between clock coordination and the agreement underlying rule following points towards a more substantive idea in the philosophy of metrology, to wit, that measuring can, and perhaps should, be analysed as an instance of rule following. Wittgenstein conceptualizes the role of standards in rule following terms. This much seems obvious given the following exchange in a 1939 seminar with Alan Turing:

> *Wittgenstein*: Making this picture of so-and-so's experiment and depositing it in the archives – you might call it doing it an honour. ...
> *Turing*: ... and when I do a multiplication ... not in your archives ...?
> *Wittgenstein*: ... We have the metre rod in the archives. Do we also have an account of how the metre rod is to be compared ...? Couldn't there be in the archives rules for using these rules one used? Could this go on forever? ... we might put into the archives just one ... paradigm ...[23]

The general point – that measuring is a rule-guided practice – is hardly worth stating. But these questions are not trivial: which account of rule following is the correct one? Which interpretation of Wittgenstein on rule following should be adopted? And what difference does the correct account of rule following make to our understanding of metrology and its investigation?

This is not the place to offer a critical review of different accounts of rule following – I have tried to provide such review in my *A Sceptical Guide to Meaning and Rules*.[24] There I defend a 'communitarian-finitist solution', building on Kripke[25] and David Bloor.[26] Recall the two main divisions in the literature on the rule following considerations. 'Individualist' and 'communitarian' renderings are divided by the question: can rule following be understood *without* or only *with* reference to a community? And 'meaning-determinist' and 'finitist' readings are separated by the question: is rule following caused by determinate mental states of meaning something by sign, or is rule following the socially sanctioned extension of an analogy with a set of exemplars?

It seems to me that Wittgenstein's metrological metaphors provide some additional evidence for the communitarian and finitist reading. Consider that for rulers and clocks to be rigid (i.e., reliable) is for them to be continuously updated and calibrated against each other. This invites a parallel idea in the case of rules: to follow a rule is a communal practice, and acts of rule following stand to rules as individual rulers stand to the communal practice of measurement. The point is reinforced by the following passage in which Wittgenstein invokes the concepts of 'honour', 'dignity', 'office', 'archive' and 'institution' when reflecting on rules or metrological units:

> The rule qua rule is detached, it stands as it were alone in its glory; although what gives it importance is the facts of daily experience.
> What I have to do is something like describing the office of a king; – in doing which I must never fall into the error of explaining the kingly dignity by the king's usefulness, but I must leave neither his usefulness nor his dignity out of account.[27]

Although this passage does not talk specifically about measuring, it occurs in a context where the relationship between fixing units of measurements and proof is the central issue.

A finitist interpretation of rule following might also draw support from Wittgenstein's proposals on what has to be deposited in the archive of indubitable mathematical propositions in order for our mathematical practice to function properly. As we saw, Wittgenstein thinks that it might be possible – under some circumstances – to deposit just one multiplication as the central paradigm. This suffices when there is sufficient agreement on how to extend the practice of multiplication from the finite set – here the unit set – of examplars:

> ... we might put into the archives just one multiplication – as a paradigm for the technique. As we might keep a paradigm of pure colour. It would make sense to do this if everyone knew from it how to multiply in other cases.[28]

But what difference does it make if we think of metrological rules along the lines of a communitarian-finitist rendering? My answer is the reminder that the finitist-communitarian reading of rule following is a central element in the Edinburgh-style 'Sociology of Scientific Knowledge'. Barry Barnes, David Bloor and John Henry spell out the implications of a finitist position for meaning and beliefs in five claims. Following their general gist but replacing 'meaning' or 'belief' with 'measurement' results in these theses:[29]

1   The future applications of metrological rules is open ended.
2   No act of measuring is ever indefeasibly correct.
3   The results of all acts of measuring are revisable.
4   Successive applications of metrological rules are not independent.
5   The applications of different metrological rules are not independent of
    each other.

If this is the correct account of metrological rules, then there obviously is scope and need for a sociological study of their functioning. In the next section I want to strengthen this link by arguing that Wittgenstein leans towards a 'metrological relativism'.

Metrological Relativism

In order to make my case, I need what I call the 'standard model of relativism'. Of course this is not an eternal or universally accepted standard. 'Standard model of relativism' merely seeks to capture some currently popular characterizations of variants of the position. I have arrived at this model by collecting definitions and characterizations of relativism from both friends and foes of the view, including Barry Barnes and David Bloor, Paul Boghossian, Gilbert Harman, Gideon Rosen, F. F. Schmitt, Bernard Williams and Michael Williams.[30] Wittgenstein's texts were not consulted. My aim is to have an independent and stable standard against which to measure his views. Finally, the suggested model could of course be developed at much greater length than I have space for here. I shall take up this challenge elsewhere. But I hope that even in its current sketchy form the model can be used as a grid or foil for understanding Wittgenstein's position.

The model has nine elements; 'Dependence', 'Plurality', 'Exclusiveness', 'Notional Confrontation' and 'Symmetry' are essential; 'Contingency', 'Underdetermination', 'Groundlessness' and 'Tolerance' are optional, though they occur in most known form of relativism.

The first two ingredients are 'Dependence' and 'Plurality':

(i) *Dependence*: A belief has a status of kind X only relative to an X-system or
     practice (=S/P). (X is a variable for elements of the set 'epistemic', 'moral',
     'metrological', etc.)[31]

(ii) *Plurality*: There are, have been, or could be, more than one such S/Ps.

Evidence that Wittgenstein accepts *Dependence* and *Plurality* are passages where he speaks of alternative practices of measuring – e.g., measuring the value of wood by taking into account only two dimensions of the piles;[32] where he writes of different ideals of accuracy;[33] where he expresses an interest in 'a plurality of self-contained systems',[34] or where he finds it possible to imagine a human society in which 'measuring quite in our sense' does not exist: 'But can't we imagine a human society in which calculating quite in our sense does not exist, any more than measuring quite in our sense? – Yes'.[35]

Wittgenstein occasionally addresses the following specific concern about *Pluralism*. Assume we say that there are, or could be, alternatives to our ways of measuring. One example of such alternatives might be the practice of determining the length of an object by means of elastic rulers. But, Wittgenstein asks, why should we think that this is an alternative way *of measuring*? Why not say that such practice is not really measuring at all? Wittgenstein replies as follows:

> It can be said: What is here called 'measuring' and 'length' ... is something different from what we call those things.
>   The use of these words is different from ours; but it is akin to it; and we too use these words in a variety of ways.
>   (Cf. pacing out.)[36]

The upshot is that the plurality of metrological practices is not ruled out by semantic considerations concerning the word 'measuring'.

(iii) *Exclusiveness*: (Two) S/Ps are exclusive of one another. This can take two forms:

   a. *Question-Centred*: There is some important yes/no question to which they give opposite answers.
   b. *Practice-Centred*: The consequences of one S/P include actions that are incompatible with the consequences of other S/Ps.

Exclusiveness tries to capture the sense in which – under a relativistic conception of their relationship – *SPs* have to *conflict*. This idea is in tension with the further assumption, made by some authors, that relativism concerns incommensurable *SPs* (here such incommensurability involves differences in categories that rule out an identity of propositional content across these *SPs*). The option of *Practice-Centered Exclusiveness* covers this eventuality. Two *SPs* can be compared, and can conflict, when they lead to, or require, incompatible forms of action and behaviour in an at least roughly specifiable area of human affairs. The requirement that the area of human affairs be specifiable safeguards that there is a certain degree of comparability. And the demand that the forms of action and behaviour involved are incompatible, makes sure that the condition of conflict is met.

That Wittgenstein adopts *Exclusiveness* can be seen from the facts that the different measurement regimes he considers – e.g., that of the odd wood-sellers or that with elastic rulers – do give answers that are different from, and incompatible with, the answers determined by our ordinary practices. However, Wittgenstein seems to be primarily interested in cases of practice-centred exclusiveness. In such cases we do not just have different answers to the same questions; we have altogether different questions. The relationship between our measuring practices and the measuring practice of the odd wood-sellers is best thought of as a case of practice-centred exclusiveness. In a different context Wittgenstein describes situations of practice-centred exclusiveness as situations where 'what interests us would not interest them'. He goes on: 'This is the only way in which essentially different concepts are imaginable'.[37]

(iv) *Notional Confrontation*: Given two S/Ps (S/P$_1$ and S/P$_2$) and a group G (that holds S/P$_1$): It is not possible for G to go over to S/P$_2$ on the basis of a neutral rational comparison between S/P$_1$ and S/P$_2$. The switch from one S/P to another has the character of a 'conversion'.[38]

The quotation given at the end of the last paragraph – 'what interests us would not interest them' – can also serve to show that Wittgenstein accepts *Notional Confrontation* for certain metrological practices as well. The tribe that relies on elastic rulers, and the tribe that insists on rulers being maximally rigid, have different ideals of accuracy. And there need not be rationally compelling arguments that lead the elastic ruler tribe to accept the rigid ruler practice. What interests the one does not necessarily interest the other.

(v) *Symmetry*:

a. *Methodological*: All S/Ps are on a par vis-à-vis a sociological / anthropological explanation.

b. *Equality*: All S/Ps are equally correct.

c. *Non-Neutrality*: There is no neutral way of evaluating S/Ps.

d. *Non-Appraisal*: For a reflective person the question of appraisal of (at least some) S/Ps does not arise.

The best-known version of *Methodological Symmetry* is perhaps the 'Symmetry' or 'Equivalence Postulate' of the 'Strong Programme' in the 'Sociology of Scientific Knowledge': 'all beliefs are on a par with one another with respect to the causes of their credibility'.[39] I generalize this 'postulate' in order to detach it from the requirement that explanations must be causal. *Non-Neutrality* is the main consideration usually invoked in defence of *Symmetry*. It does not preclude the possibility that some *SP*s agree on the standards by which their overall success should be judged. What *Non-Neutrality* denies is that such local agreement justifies the hope for a global or universal agreement. Most characterizations of relativism – by friends and foes alike – take *Equality* to be the natural consequence of *Non-Neutrality* and thus the best way to spell out *Symmetry*. But *Equality* makes a stronger claim than *Non-Neutrality*. This becomes easy to appreciate once we remember the typical challenge to *Equality*: what is the point of view from which *Equality* is asserted? On the face of it, *Equality* appears to presuppose a neutral point of view from which we can somehow see that all *SP*s are equally correct. And this very claim jars with *Non-Neutrality*. *Non-Appraisal* seems to avoid the problems of *Equality*, while capturing the important core of *Non-Neutrality*. It is motivated by the thought of 'intellectual distance': the idea that a reflective person holding one *SP* might come to the conclusion that her own 'vocabulary of appraisal' simply does not get a proper grip on the judgements and actions of another *SP*. It is not that this vocabulary could not possibly be applied at all – it is rather that such application seems forced, artificial and contrived.[40]

Wittgenstein never formulates the idea of methodological symmetry clearly, though it seems to me that many of his discussions, especially his criticisms of the anthropological work by Frazer, in fact rely on something like it.[41] Since I cannot go over these issues here, I must confine myself to drawing attention to two quotes. In the first Wittgenstein rejects 'reason' as an asymmetrical principle for analysing how different people in different contexts justify claims about causality.

> Reason – I feel like saying – presents itself to us as the gauge par excellence ... This yardstick rivets our attention and keeps distracting us from these phenomena...[42]

The second passage offers a sociological-historical explanation for why the odd wood-sellers calculate the value of wood as they do. There is no suggestion here that this type of explanation is only for 'the deviants'. In fact Wittgenstein is adamant that it is not restricted in this way:

> (a) These people don't live by selling wood ... (b) A great king long ago told them to reckon the price of wood by measuring just two dimensions (c) They have done so ever since ... Then what is wrong? They do this.[43]

At some points in the 1930s Wittgenstein seems to have been tempted by *Equality*, writing for instance that 'one symbolism is as good as the next'.[44] And about the comparison between finite and infinite systems of primes he says that the latter is with 'no greater rights than the former'.[45] In 1942, however, Wittgenstein rejects *Equality* as meaningless, at least for the case of ethics: 'If you say that there are various systems of ethics you are not saying they are all equally right. That means nothing'.[46]

Non-Neutrality regarding metrological standards is most clearly alluded to in a comparison with ethics. At issue is the choice between Nietzschean and Christian ethics:

> ... suppose I say Christian ethics is the right one. ... It amounts to adopting Christian ethics.... surely one of the two answers must be the right one. ... But we do not know what this decision would be like ... Compare saying that it must be possible to decide which of two standards of accuracy is the right one.[47]

Just as there is no neutral way of deciding between standards of accuracy, so there also is no neutral way of deciding between ethical systems.

Non-Appraisal surfaces most clearly in Wittgenstein's thoughts on ethics: 'Has a man a right to let himself be put to death for the truth? ... For me this is not even a problem. I don't know what it would be like to let one-self be put to death for the truth'.[48] Nevertheless, the ways in which Wittgenstein discusses different metrological regimes show that he find *Non-Appraisal* natural in these contexts, too:

To call something 'inaccurate' is to criticise it; and to call something 'accurate' is to praise it. And that means: the inaccurate does not reach the aim as well as the accurate. But here all de-pends what we call 'the aim'. ... There is not just one ideal of accuracy.[49]

And concerning the odd wood-sellers we are asked: 'But there is nothing wrong with giving wood away. ... Is there a point to everything we do'?[50]

(vi) *Contingency*: Which S/P a group G finds compelling is ultimately a question of historical contingency.

(vii) *Groundlessness*: There can be no X-type justification of S/Ps.

Contingency and *Groundlessness* are general underlying assumptions of Wittgenstein's later work. There is no reason to suspect that he would exempt metrological issues from their domain. Two well-known passages are worth citing:

And how can I know, what – provided I lived ... completely differently – would seem to me to be the only acceptable picture of the world-order?[51]

What has to be accepted, the given, is – so one could say – forms of life.[52]

The last two elements of my 'standard model' of relativism are *Underdetermination* and *Tolerance*:

(viii) *Underdetermination*: S/Ps are not determined by specific facts of nature.

(ix) *Tolerance*: At least some S/Ps other than one's own must be tolerated.

Suffice it here to illustrate Wittgenstein's adherence to *Underdetermination*. *Tolerance* has been amply documented in the above. I have already quoted the following passage above: '...the rules of grammar are arbitrary in the same sense, or non-arbitrary in the same sense, as the choice of unit of measurement'.[53] The degree of this latter arbitrariness is limited: there are 'general facts' that make certain types of measurement salient and practical. Note, however, that Wittgenstein insists on the ultimate source of normativity being not these general facts but human communities. While there *are* 'general facts ... that make measurement with a yard-stick easy and useful ... it is *we* that are inexorable' in insisting on its continuing use.[54] In another place Wittgenstein explains what kinds of general facts he had in mind:

'There are 60 seconds to a minute'. ... could we talk about minutes and hours, if we had no sense of time; if there were no clocks, or could be none for physical reasons; if there did not exist all the connexions that give our measures of time meaning and importance? In that case – we should say – the measure of time would have lost its meaning...[55]

Or, for another example, in his various writings about colour classification and measurement, Wittgenstein mentions the following general facts: in our natural environment, colours are not tied to specific forms; we have the technologies to produce dyes and colour things; colours are not always linked to specific smells or threats (poison); no one colour dominates our environments; we are very skilled at producing blends out of primary (and secondary) colours; for instance, 'reddish yellow' out of red and yellow; and the other qualities of a thing are connected with its colours: hence grass is green, chalk white, blood red.[56]

To sum up this lengthy section: Wittgenstein leans towards a form of metrological relativism. He also seems to adopt the methodological *Symmetry Principle* central to the 'Strong Programme' of the Sociology of Scientific Knowledge. These features strengthen the suggestion that there is an affinity between Wittgenstein's reflections on metrology and the Sociology of Scientific Knowledge.

Van Fraassen, Wittgenstein and the Problem of Coordination

I finally turn to my attempt to relate Wittgenstein's remarks on metrology to the recent work by Bas van Fraassen. To repeat, van Fraassen's aim is to develop and defend a 'constructive-empiricist form of structuralism', that is, a form of structuralism that leaves room for agency, indexicality and contingency.

Van Fraassen offers the following general account of measurement: 'The act of measurement is an act – performed in accordance with certain operational rules – of locating an item in a logical space'.[57] Van Fraassen suggests that Wittgenstein's *Tractactus* contains the first clear formulation of this idea; for instance in 2.013: 'Every thing is, as it were, in a space of possible atomic facts...' One implication of this account is that measurement need not involve numerical scales.[58] More importantly, theories are (like) instruments in that they too locate items in a logical space in accordance with certain operational rules.

Van Fraassen's interpretation of the local and the global problems of coordination goes as follows. Local coordination concerns measuring instruments as ordinarily understood. The problem is this: How do we find and justify a function 'f' that adequately maps states of a given measuring instrument to physical magnitudes? More precisely, how can we do so, without already using the results of measuring with this very measuring instrument? If our only access to the physical magnitude is this very measuring instrument then obviously the problem cannot be solved.

The problem of global coordination relates to theories or languages as 'measuring reality'. In this context van Fraassen relies on a version of Putnam's famous 'model-theoretic argument': which theory should we accept given that all consistent theories come out true on some interpretation (f_i) of their predicates? There is no solution to this problem taken abstractly.

Neither the local problem nor the global problem can be solved relying on the 'view from nowhere' offered by forms of 'pure structuralism'.[59] These are forms of structuralism that bracket history and human agency. Solutions are available, however, once we allow history and human agency back into the picture. Thus the local problem of coordination is usually solved 'from within' a historical situation in which other measurement procedures are contingently available.[60] And the global problem disappears once we focus on the fact that we 'always already' have a language with its interpretation – an interpretation that we cannot step out of. Van Fraassen concludes:

> The response is Wittgensteinian, in that it focuses on us, on our use of theories and representations, and brings to light the impasses we reach when we abstract obliviously from use to use-independent concepts.[61]

These ideas can now be brought into contact with the interpretation of Wittgenstein on metrology suggested above. To begin with, note that Wittgenstein sees both the local and on the global versions of the problem of coordination – at least in a somewhat cryptic form:

> ...is the unit of measurement ... the result of measurements? Yes and no. It is not the result of measurements but perhaps their consequence (Folge).[62]

> The rules of grammar cannot be justified by showing that their application makes a representation agree with reality. For this justification would itself have to describe what is represented.[63]

Moreover, van Fraassen is right to regard Wittgenstein's thought as central for identifying the solution to the global problem of coordination. Nevertheless, van Fraassen's position remains unsatisfactory without an explicit reference to a specific rendering of the rule following considerations. In other words, 'use', 'intention' and 'convention' are not by themselves solutions to the problem of coordination. We also need to understand in an appropriate way these concepts and the phenomena in their extensions. That is to say, we need to understand them in communitarian-finitist ways – or else the skeptical paradoxes simply reappear on the level of the subject.[64]

Finally, van Fraassen is combating the de-contextualisation of structure in pure structuralism and its measurement theory. This is as it should be – from a Wittgensteinian perspective. But Wittgenstein offers more than just a general support from van Fraassen's starting point. Wittgenstein's (admittedly rudimentary) sociological analyses of units of measurement and their analogues – rules of grammar, proofs, certainties – help us understand why the decontextualization or mystification of structures is hard to avoid. Consider once more the key passage I cited earlier:

> The rule qua rule is detached, it stands as it were alone in its glory; although what gives it importance is the facts of daily experience.

> What I have to do is something like describing the office of a king; – in doing which I must never fall into the error of explaining the kingly dignity by the king's usefulness, but I must leave neither his usefulness nor his dignity out of account.[65]

Rules, standards or structures have authority precisely because and insofar as they have dignity. And such dignity is often based precisely on the suppression of involvement with practice, subjectivity and agency. On a general level this has long been recognized in sociology.[66] Wittgenstein makes this insight fruitful for the sociology of metrological knowledge.

Conclusions

Wittgenstein's importance for the sociology of scientific knowledge has of course been emphasized before. In this paper I have tried to give substance to a more specific thesis, to wit, that Wittgenstein offers important building blocks for a sociology of metrological knowledge. And I have sought to make plausible that this form of metrology might constitute a crucial ally for van Fraassen's criticism of pure structuralism.

2 METROLOGY AND VARIETIES OF MAKING IN SYNTHETIC BIOLOGY

Pablo Schyfter

Introduction

In *The Logic of Modern Physics*, P. W. Bridgman proposes an understanding of metrological concepts based on the operations necessary to produce measures. A metrological concept (such as length) has meaning because and *only* because there exists a 'set of operations by which [the measure] is determined'[1] (such as laying measuring rods end to end or using the reflection of a laser). The concept and the operations are one and the same; a change in operations results in a change in concept (thus, rod length and laser length are not the same thing). This perspective shares much with J. L. Austin's view on certain linguistic acts;[2] simply stated, measurement operations are performative of measurement concepts. In this volume, Martin Kusch examines Wittgenstein's views on measurement to similar effect[3] – acts of measurement define and sustain concepts of measurement. As a result, Kusch argues, communitarian epistemology and the sociology of knowledge are useful perspectives with which to examine metrology. Here, I use a case study from synthetic biology to demonstrate that metrological practices make more than just concepts. They also define fields, their members and their objects.

Bridgman's operationalism[4] examines how acts and concepts of measurement come into being together. My study demonstrates that the consolidation of a field, the making of uniform objects, and the normalization of metrological practices are different facets of a single phenomenon: the making of a metrological community. Using communitarian epistemology and original empirical research, I establish two points. First, to understand metrology is to understand a metrological community. Second, to understand that community is to understand how its practitioners, their work and their objects are ordered. Thus: to understand metrology in synthetic biology is to understand a community dedicated to proper engineering; to understand that community is to understand an engineering-based ordering of the collective, its objects and its participants' acts.

The Study

The empirical material I present here is the result of an eighteen-month ethnographic investigation of a synthetic biology laboratory in the United States. During that time, I worked alongside the unit's researchers and interacted formally and informally with them. I carried out interviews with individuals from my host lab and its partner lab, which shared space, equipment and some personnel with my own. I also studied other laboratories at seven different institutions in the United States through short-term ethnographic research and in-depth qualitative interviews. In total, I carried out twenty-four such interviews. Moreover, I attended events central to synthetic biology, including: the Fifth and Sixth International Meetings on Synthetic Biology; research networks meetings such as those for the US Synthetic Biology Engineering Research Centre; and the yearly undergraduate International Genetically Engineered Machine competition.

Synthetic Biology and Engineering: Consolidating the Field

I begin by describing the relationship between synthetic biology and engineering. Synthetic biology is an unsettled, heterogeneous field.[5] At present, it is not characterized by any single identity, nor does it have definitive practices, concepts and goals[6] – in many ways it is 'all things to all people'.[7] Those who self-identify as synthetic biologists come from a range of disciplines and employ a variety of ideas and methods to achieve a diversity of aims. Nonetheless, factions exist within this community that display greater coherence. The most prominent of these, and the one I explore here, wants to make synthetic biology into an 'authentic' engineering field: to consolidate the community around a shared set of engineering principles, practices, aims and expectations.

Members of the engineering contingent have published a number of review articles in which their broad mission – the consolidation of the field using an engineering mould – is expressly articulated. For instance, Endy argues that the basic target of the field is to 'make the engineering of biology easier'.[8] Similarly, Heinemann and Panke claim that synthetic biology is about 'putting engineering into biology'.[9] My interviewees expressed similar views.

For many of those I interviewed, engineering (as target and as practice) is vital to the identity of synthetic biology. Karen[10] is a principal investigator and dedicated champion of synthetic biology. Her laboratory's research aims at advancing the engineering capacities of synthetic biology. Some of the research seeks to produce so-called foundational technologies – tools to facilitate future work along engineering lines. When I asked her to describe synthetic biology, she replied:

... I, first of all go back to how we defined it ten years ago. Which was a group of computer scientists and bioengineers, asking a simple question, which was, 'how can we make biology easier and more predictable to engineer'? And that was the basis for forming synthetic biology. As far as I am concerned, end of story.

As do the authors I mention above, Karen describes synthetic biology as an engineering field with engineering goals and an engineering pedigree. For her, engineering is so vital to the identity of the field that she reduces all of synthetic biology to its engineering target ('end of story'). Karen also explained the importance of 'authentic' engineering as a mechanism for differentiating the field from the already-existent genetic engineering. Setting and securing synthetic biology's identity depends upon establishing such a boundary[11] – a border defined by engineering.

Members of synthetic biology's engineering contingent routinely portray genetic engineering as not authentic engineering, unlike their own field. Gary, another principal investigator, argues that the field will practice and produce as do disciplines like civil, electrical and mechanical engineering. His laboratory makes basic tools and techniques for synthetic biology: mechanisms that enable the rapid assembly of small segments of DNA and reduce the unpredictability of functional constructs, and conceptual tools that make the design of biological technologies easier to carry out. He has argued in publications that synthetic biology will replace the craftwork of genetic engineering with the systematic rigour of established engineering. When interviewed, he made this same point:

So I think, there's that idea of putting the engineering back into genetic engineering which is, you know, turning it into something very formal, theoretical, having a theoretical frame for how to do it. I think, that is a very real aim of the field.

Gary's view that synthetic biologists must follow a particular type of conceptual scheme and carry out specific forms of practice is shared by those who sponsor the engineering view of the field. It is also the primary method by which practitioners draw boundaries between their work and that of genetic engineering.

Barry, whose work on mechanisms for failure prevention in biological technologies is based in great part on systems and control engineering, also articulated a sharp distinction between synthetic biology and genetic engineering. For him, the latter has too little engineering:

They call it genetic engineering, but for example, they are not necessarily engineers doing that, they are a scientist ... [synthetic biology] is bringing a lot of engineering to the table of what we can do with biological circuits. Just the word 'circuit', is an engineering concept, it is not something they are going to actually use in traditional biology.

Science and engineering are distinguished by an important boundary, Barry argues. To justifiably lay claim to the identity of 'synthetic biologist', a practitioner must position herself on the appropriate side of that boundary. Michael, a postgraduate student at my host laboratory, expressed a similar view. The transition from genetic engineering to synthetic biology, he states, involves moving away from science. Though the field begins 'with people just doing something based on science 100%', Michael says, practitioners will with time 'know more and more engineering principles', which will enable proper engineering practice and through it, the making of a proper engineering field.

The engineering ideal also serves as a normative guidepost for individuals within the community. Practitioners must position themselves and behave in ways that satisfy engineering expectations; valid members of the community are those who demonstrate engineering praxis and ethos. Robert, a principal investigator who loudly advocates the engineering vision, argues that what 'distinguishes synthetic biology from the broader field of biotechnology ... is that it does a better job of recognizing what a great engineer is all about'. That is, synthetic biology employs normative criteria common to traditional engineering. Important here is the production of enabling technologies, Robert says:

> A great engineer doesn't just deliver a technology that solves a particular application ... a great engineer will develop tools that make the solving of all problems of that similar category, the same category, easier ... a great engineer will deliver the application, but actually, will make it easier to deliver on all sorts of applications in the future.

For Robert, the quality of a practitioner in synthetic biology depends upon her willingness and ability to advance the field *as an engineering discipline*. For him, as for others in the engineering contingent, this involves delivering foundational technologies. Crucially, individuals' behaviour and membership is to be evaluated following an engineering ideal. Practitioners are rewarded and sanctioned according to engineering norms.

In summary, synthetic biology is to be proper engineering and its practitioners are to be proper engineers. The field's boundaries are set according to conceptions of engineering. Its members' behaviour is to be evaluated according to engineering norms. The field and its members' practices are to be ordered in an engineering fashion.

Synthetic Biological Parts: Making Objects Uniform

For advocates of the engineering ideal, making synthetic biology an authentic discipline of engineering involves nothing more than enabling authentic engineering practice: Heinemann and Panke argue the two are 'synonymous'.[12] This latter goal demands a series of so-called foundational technologies: tools that can 'make the design and construction of engineered biological systems easier'.[13] Of these, the making of modules for construction – synthetic biological parts – has received considerable attention.[14] Such modules are to be the nuts and bolts of synthetic biological artefacts. They are to be physically and functionally uniform and their use is to be routinized.[15] Importantly, the objects are to satisfy an engineering order.

Because being engineering means *doing as* engineering, the making of standardized parts is involved in the setting of boundaries and the demarcation of synthetic biology as something other than genetic engineering. Juliet, a renowned principal investigator, oversees a range of projects all aiming to deliver new, engineering-focused tools and techniques, including foundational technologies. When asked about the identity of synthetic biology, she replied:

> So, to me, that specific sort of, focus on the application of engineering principles, and the development of sort of, foundational tools, is something that really sets synthetic biology apart ...

For her, the making of enabling technologies defines the field. A similar, but more specific argument was made by Andrew, another prominent synthetic biologist. More than others quoted here, he pursues the making of products for commercial ends. To his chagrin, his research is often classed as genetic engineering despite his own self-identification with synthetic biology. During the interview, he argued vehemently for his place in synthetic biology by emphasizing the importance of and his commitment to standardized parts. He said:

> ... genetic engineers or the molecular biologists, I think, don't think about modular parts and standardised parts, and assembly of parts ... I think the mindset is different. And so, it is about the engineering, for me.

Andrew points to how practice distinguishes fields and establishes disciplinary identities. In so far as engineers practice differently from scientists,[16] the making of uniform parts serves to set boundaries between what synthetic biology hopes to be and what it sees genetic engineering as being right now. Being engineering is *doing as* engineering and *having as* engineering: producing and having access to standardized parts.

Crucially, parts are to be characterized. According to their advocates, synthetic biological parts are to display "plug-and-play" compatibility'.[17] That is, parts are to feature reproducible and robust functionality – predictable utility. Enabling such performance demands making parts of a kind uniform and producing reliable descriptions of each kind. Parts must be rendered physically and functionally standard and then characterized using shared descriptors. Without achieving both aims, users would be unable to choose appropriate parts for specific needs, nor could they expect predictable, reliable performance: parts would not be uniform. Fulfilling these two goals still presents sizable difficulties. Rose, a doctoral student at Karen's laboratory, argues:

> … as much as people in the field like to talk about, parts are characterised and they are going to be standard, and so we can share them, and you'll know how it works because I measured it, and then you'll know how it's going to work in your system. One of the difficulties is that, that type information that those quantitative descriptions, that just doesn't exist for most parts. And even if it does exist for a part, more often than not, the way that it was quantitatively described or characterised, would not be applicable to how you're going to use the part.

Parts are not sufficiently standardized, nor are the methods for satisfactory quantitative characterization yet in place. Frank, a principal investigator whose laboratory works on tools for measuring, modelling and analysing living systems, concurs: 'everyone in synthetic biology agrees that we are not characterizing things well enough'. Data sheets reporting the behaviour of individual parts do not exist for all parts. Attempts to produce a central database for such information (or even the parts themselves) have not succeeded as hoped. Last, methods for developing quantitative measures – metrological practices – have not been adequately normalized.

An example of parts characterization in synthetic biology may serve to illustrate the relationship between characterization, quantification and metrology. Research at the BIOFAB, a synthetic biology centre in California, has attempted to standardize and characterize transcriptional promoters.[18] Promoters are genetic sequences that trigger and drive the transcription and translation of other sequences. In a sense, they are parts that operate as on/off switches. The BIOFAB's characterization of commonly used promoters employed fluorescent reporters in order to measure the intensity with which different promoters drive transcription and translation. Promoters were paired with a gene that triggers the making of a fluorescent protein. 'Stronger' promoters cause more of the protein to be produced; thus, by measuring the fluorescent intensity of a colony of organisms transformed to carry the promoter-fluorescence construct, researchers can infer the strength of each promoter. A quantitative measurement of fluorescence helps practitioners characterize promoters according to each part's strength. *Quantification* of fluorescence, a practice in *measurement*, supports a particular (and here the preferred) type of *characterization*. Moreover, these three practices contribute to the ordering of parts: such-and-such a promoter is stronger than another.

In summary, the objects of synthetic biology are to be made uniform as objects *of proper engineering*. Their design, construction, evaluation and use are to be ordered in an engineering fashion. A crucial aspect of this is quantitative characterization – a practice that requires normalized engineering metrology.

Bringing Measurement to Life: Normalizing Metrology

Rose and Frank's frustration with missing metrology is shared by many in synthetic biology. For these researchers, making uniforms parts and consolidating the field as one in engineering demand normalizing metrology: developing standard measurement operations and making their use routine.

Recall that constructing synthetic biology's engineering identity involves distinguishing it from efforts such as genetic engineering. Karen, quoted above, argues that such research in biotechnology has been 'a little non-quantitative and a little bit *ad hoc*'. For her, the two flaws must be resolved together: enabling systematic design and construction requires quantitative measurement. Frank makes a similar point:

> I don't think it is so hard to build something, it's hard to build something that you know works or, again, because we probably just have such bad assays. And we don't even know so much what we are trying to measure.

Making technological artefacts is to be a defining feature of the field as one in engineering. Importantly, it is a pursuit that demands shared metrology. To do what it wants to do and be what it wants to be, the field must come together in measurement.

My interviewees presented a number of reasons for the current lack of metrological capacity in synthetic biology. For instance, Frank pointed to the recalcitrant and unpredictable character of living nature. More importantly, practitioners discussed the absence of metrological standards. Andrew, quoted above, says:

> So, say you do fluorescent measurement on this machine, you go to the next machine that is sitting right beside it, it could be very different. You could go to your colleague, a few cities away, and do the measurement, even more different, right?

Metrological contingency is a serious challenge to synthetic biology, particularly when there are no set ways of carrying out particular measurements. Juliet says:

> ... people all sort of, have their favourite way of, you know, measuring something. You know, and maybe they don't have the exact same sort of, measuring devices in their laboratory, right? Meaning, some people might have a very nice, sort of, right, flow cytometer that allows them to do multi-colour analysis ... Other people might only have like, a luminometer, right, or something to do beta-gal ... So that can make it hard, right, to sort of, standardize measurement, so require the people to measure things in a specific way. It is also the case that, even sometimes, the set-up of these instruments is different because you set them up in different ways, and you set gains and all this other stuff in different ways, right?

Local variations in equipment, preferred methods, requirements and aims result in a lack of metrological consistency. Measuring a basic characteristic of a synthetic biological part, such as the activity of a transcriptional promoter, can be done in a number of ways, each of which requires a different set of procedures and instruments. At present, no single method has a definitive lead over the rest, much less is any the established method for the field. Lack of consensus is the problem.

The need for metrological consensus came up repeatedly, often as a crucial hurdle for the field. Karen made this point when I asked about synthetic biology's metrological capacity:

> We can measure lots of things. I mean, we can measure single molecule levels in cells. We are really good at measurement ... So, ways to measure things are always, that's a big part of technology right now. So, that's a very rapidly moving area, there will be no worries there. The question is, do we actually need all those measurements?

Karen argues that the capacity for precise and useful measurements exists, but there is too much variety in methodology. She identifies a fundamental problem: too much is measured because there is a lack of agreement about what to measure. Not enough attention is paid to what kinds of measuring are necessary, which are not so, and which necessary kinds are currently unavailable. Advancing the field as one in engineering demands communal answers to these queries.[19] The problem is not one in technical capacity, but one in collective agreement.

Unfortunately, collective agreement – consensus about what and how to measure – is lacking. Nick, a postdoctoral researcher at my host laboratory, argued that agreement is both absent and necessary. He said:

> ... in terms of standardization of measurement, it's a shambles, I would say, right now. There needs, there definitely could be more standardization ... And it is going to require, you know, organization, because it is a lot harder. You know, it's a lot harder to be strict about how you measure things, the way you measure it. A lot of biologists, I think their tendency is to just get it done and not worry about, you know, how, somebody else can be able to compare my numbers, you know, five years down the road.

Nick points to the lack of metrological consensus and the importance of communal organization in order to enable agreement. Conventional practice in biology, he argues, does not encourage (much less demand) the type of metrological consistency that is common to established engineering disciplines. Simply stated, metrology is to play an important role in establishing synthetic biology as a field in engineering and one distinct from biological sciences. Its disciplinary character and metrology are entwined.

Other interviewees pointed to important structural difficulties that lead to metrological inconsistency and a lack of work to remedy this fault. Gary, Frank and Andrew all suggested that the field lacks practitioners willing to perform metrological research because such work is undervalued by the broader scientific community. Frank argued that without 'a link between high quality characterization of parts and getting a paper published in *Nature*, the field may not care as much'. He asks, 'where do you publish datasheets? How do you get academic credit for making datasheets'? Simply stated, it is unlikely that practitioners will carry out difficult work that goes unrewarded or does not lead to professional advancement. Lucy, a doctoral researcher at Karen's laboratory, argued similarly. Parts characterization and measurement are vital to the development of synthetic biology, but achievements in this domain are 'not a good PhD thesis anymore'. The research is not encouraged or admired. Thus a structural feature of the field – metrological research is not regarded highly or rewarded well – discourages important, necessary work.

A second structural challenge is important. At present, no central facilities exist to construct, characterize and validate parts. Whenever a component is borrowed, the borrowing laboratory must repeat key measurements anew. The alternative, argues Frank, is to 'rely on what some other lab may have said, which may or may not be reliable'. Metrological inconsistency and the lack of broadly respected providers means that one laboratory cannot trust the parts or measures produced by another. The objects intended to be the nuts and bolts of biological engineering cannot be used as are nuts and bolts in mechanical engineering. Rose, the doctoral student quoted above, says that 'something that is fairly frustrating is that, you often have to characterise everything yourself, at this point'. Central facilities and shared providers will not resolve all of these issues instantly. 'There will always be trial-and-error', Lucy argues, but 'it would be great to ... start with something that's a little better understood'. Put otherwise, it would be great to practice as do established engineering disciplines: with no lack of 'trial-and-error'[20] but with much material and metrological consistency. Frank, Rose and Lucy are noting a lack of trust in the reliability of measurements and the performance of parts. Again, these problems have everything to do with lack of communal agreement. While technical challenges exist to making parts and measures uniform, the greater challenge is to do with consensus.

In summary, metrological practices are both necessary and currently unsatisfactory in synthetic biology. The key problem is the lack of metrological consensus in the community: measurement needs to be normalized and coordinated collectively.

Making Measures, Making Things, Making Fields

The three preceding sections have established three claims, respectively. Synthetic biology intends to consolidate itself as a field of engineering. It intends to produce uniformly engineered objects. It intends to normalize practices of engineering metrology. These three aims are facets of a broader process: the making of a metrological community.

Martin Kusch, in his comprehensive presentation of and argument for communitarian epistemology, writes:

> To understand knowledge is to understand epistemic communities; and to understand epistemic communities is to understand their social and political structures.[21]

My argument here follows directly from this claim. *To understand metrology is to understand metrological communities.* This is the case both because metrology, like knowledge, exists only in and through social collectives, and because metrology in turn serves to order the community, its objects and its practices.

Consensus and Community in Measurement

The sociology of knowledge and communitarian epistemology stress that knowledge is a collective good. Knowledge presupposes the existence of 'a plural, communal subject, a "we"'.[22] It is because any such 'we' exists that knowledge claims can be established, justified and transmitted; it is because any such 'we' exists that correct and incorrect uses of knowledge claims can be distinguished; and it is because such 'we's exist that there can be multiple, incompatible, and internally consistent bodies of knowledge. So too is the case with metrology. In his contribution to this volume, Kusch makes this point in advocating 'metrological relativism'.[23] The validity of metrological systems is established through social consensus; any single act of measurement is correct in so far as it meets collective approval (or put otherwise, avoids sanctioning by community members); and different collectives can give rise to incompatible, internally consistent metrologies. Thus communities – and specifically, agreement within a community – enable particular systems of measurement to exist and persist.

Consider synthetic biology. This group's desire to consolidate the field around a shared set of engineering principles, practices, aims and expectations characterizes the type of metrological work these researchers seek to enable: *engineering metrology.* Were this community seeking a different disciplinary end, its metrological ambitions, concepts, practices, instrumentation and norms would be correspondingly different. So too would be what measurements are considered valid, how they are validated, and when consensus is reached or missed.

I submit that metrological consensus can also work to bring a community into being. Thus while metrologies are the result of social agreement and are shaped by the conventions of each collective, as Kusch agues, they can also contribute to bringing such communities into existence. Most broadly, this is the case because of synthetic biology's unsettled and heterogeneous character at present. If metrological consensus can be accomplished, it will serve as an act (ideally, one of several) in unifying disparate practices and practitioners. However, there exist more nuanced reasons for the potential of metrological consensus to help consolidate the field. Recall that synthetic biology is to *be* engineering by *doing as* engineering. Heinemann and Panke write that 'to be successful as an engineering discipline, synthetic biology will need to repeat the corresponding developments of its sister engineering disciplines'.[24] One such 'development' is the production of metrological tools and techniques of the kind 'widely used in the electrical, mechanical, structural and other engineering disciplines'.[25] My interviewees repeatedly identified quantitative measurement and characterization of objects as necessary capacities for making synthetic biology into a field of engineering. They also emphasized that a lack of metrological consensus is the key problem facing efforts to establish engineering measurement in synthetic biology. Without a standardized and broadly accepted system of measurement – without normalized metrology – necessary advances in this craft will not come to pass. Without this craft, attempts to establish synthetic biology as an engineering discipline will not progress.

At present, there exists no metrological consensus in the field. Should such agreement come to pass, it will assist in the making of a field that remains still in-the-making. Metrological consensus will help bootstrap the discipline into being, order that discipline once it exists, validate its status as engineering and sustain that status over time. It will not do this determinatively, nor will it do it alone, but such consensus will assuredly help the process along. The interviewees suggested a plethora of ways in which this is to happen. Bringing systematic, quantitative procedures to synthetic biology will help make rigid the boundary between this field of 'proper' engineering and existing ventures in genetic 'engineering'. Quantification of the field will advance its similarity to established engineering disciplines. Shared metrological practices – consensus – will help unify the field's currently heterogeneous population along engineering lines. This same consensus will help mould members by sanctioning individual deviation from the engineering ideal. Put otherwise, metrological consensus will act to establish, validate, order and sustain a community. *Making measures and making the field are tied together.*

Making Measurable Objects

Making sense of metrology is making sense of communities. All communities have their objects: some because they produce them, all because they have shared conventions about how to make sense of entities and phenomena. Kusch argues that there exists no knowledge of natural things 'without knowledge about communally instituted taxonomies, standards, and exemplars'.[26] In looking out my office window, I make sense of this object facing me through a classificatory system specific to my community. Because I am familiar with my community's kinds, have exemplars for each, and can be corrected by others when I stray from shared beliefs, I can make sense of the object as a 'tree'. Absent my community or if I should belong to another, the object would fall under different taxonomies, and I would make sense of it in a correspondingly different way. It is in this sense that communities have their objects. Our knowledge – a collective good particular to a community – makes objects intelligible. Importantly, Kusch tells us, this knowledge is not a detached description of things out there; beliefs 'shape their referents'.[27] Our empirical beliefs 'concern both what is and what ought to be', and thus they 'fit the world only because they also form the world'.[28] Communities engage their objects in ways particular to the beliefs that render those objects intelligible. So too is the case with measurement and metrological communities, which render, order and validate objects.

Synthetic biology hopes to be in the business of designing and fabricating technologically functional biological artefacts. Doing so involves planning and producing material entities, as well as rendering such entities intelligible in particular ways. That is, the field's objects must be *made* in the ordinary sense and in the sense of being placed within an ontological order.[29] Metrology plays a role in both.

Above, I presented practitioners' views that quantitative measurement is a key capacity for enabling synthetic biology to produce functionally uniform and operationally reliable parts *in an engineering fashion*. The researchers made similar claims regarding the technological products those parts are intended to underpin. *In engineering*: designing material technologies involves quantitative specifications of form; producing technologies involves ensuring that those specifications are met as accurately as is necessary; and evaluating produced technologies demands quantitative tests of performance. These activities all involve measurement. Moreover, the very material constitution of objects will be shaped by engineering metrology. After all, the objects to be measured must be made physically capable of being measured in the first place (say, by incorporating reporter genes). Thus, communal expectations of measurability affect both the manner in which objects are to be made and the form those objects are to take.

Measurement is also involved in rendering objects intelligible as objects of 'proper' engineering. Things are intelligible to us in particular ways because of our communally instituted beliefs. Synthetic biologists view biological things as objects of engineering: usable substrate for an enterprise in technological design and construction.[30] They are engineering and 'engineerable' objects. To bring engineering metrology to bear upon biological entities is to understand these as measurable objects. Most basically, because they are subject to the concepts and practices of a particular metrology: they can and are to be measured. More significantly, their intelligibility is structured according to quantifiable parameters of design, construction and performance. Objects can and are to be understood by way of quantitative datasheets.

Enabling intelligibility is the core, but not sole, manner in which metrological communities order their objects. Advocates of engineering in synthetic biology seek to make 'well characterized parts'.[31] Characterization involves quantitative measurement of parts' performance and will produce datasheets. Practitioners will use these to order parts according to function and behaviour: such-and-such part does X function at a rate of Y. What is measured and how it is measured will be employed to organize functional components, as happens in established engineering fields. Doing so produces pragmatic taxonomies designed to enable the design and production of technologically functional artefacts.

Last, objects are to be evaluated following metrological norms. Their validity as 'proper' objects of engineering follows from their measurability. Inferior parts are those not yet or incapable of being given quantitative characterization. Satisfactory, usable parts are those with sufficient quantitative characterization. Proper engineering produces objects that can be understood and evaluated using systematic quantitative metrology. Thus to be proper objects, synthetic biological technologies must fit this requirement.

Following Kusch, metrological beliefs and practices concern both what is and what ought to be. Synthetic biological objects are to be quantifiable entities of engineering. Their shape, the ways in which they are made, how they are to be understood, ordered, used and evaluated all reflect the metrological particularities of this community. *Making measures and making objects are tied together.*

Conclusions: Measuring and Making

Those I studied define the boundaries of synthetic biology, as well as inclusion and exclusion in the field, following a conception of engineering. For them, the field is to be consolidated following an engineering order. The objects of synthetic biology are to be made uniform as engineered and as *of* engineering. Their form and ontology are to follow an engineering order. Measurement in synthetic biology is to be normalized – homogenized and made routine – according to an engineering mould. The goal is ordering measuring practices as do established fields of engineering. All three things – field, objects and practices – are to be structured and validated as facets of 'proper' engineering.

Importantly, each variety of structuring and validation is entangled with the other two. Success in engineering metrology will allow objects to be validated as properly of engineering. Success in producing proper objects of engineering will validate the field as one authentically in engineering. Measuring contributes to the making of things, which in turn helps establish the field. Normalizing metrology, making objects uniform and consolidating the field are different facets of a single enterprise: the making of a metrological community.

This provides us with a compelling communitarian elaboration to the ideas with which this article began – arguments from P. W. Bridgman's operationalism. Bridgman proposes an understanding of metrological concepts based on the operations necessary to arrive at a measure.[32] Shared measurement operations bring a shared concept into being. My work with synthetic biologists shows that metrological consensus – collective agreement about what and how to measure – also brings into existence a metrological community. This 'we' brings with it a communal identity, expectations of its participants, and an ontological order. Making measures creates the metrological concept, but it also helps define the field, its members and its objects.

3 REFRAIN FROM STANDARDS? FRENCH, CAVEMEN AND COMPUTERS: A (SHORT) STORY OF MULTIDIMENSIONAL ANALYSIS IN FRENCH PREHISTORIC ARCHAEOLOGY

Sébastien Plutniak

In 1981, Françoise Audouze and André Leroi-Gourhan published a critical overview of the evolution and current state of prehistoric and classical archaeology in France. This paper had an explicit message, summed up in its title, 'France: A Continental Insularity'.[1] According to the authors, this insularity was mainly due to the weak theoretical developments in French archaeology compared to English-speaking archaeology. The *New Archaeology* had been flourishing in the US and the UK since the beginning of the 1960s. Pioneered by Lewis Binford,[2] the novelty of this archaeology lay in an ambitious scientific strengthening of the discipline. Controlled sampling, statistics and the use of models were considered the best ways to reach this aim, and prehistorians tried to standardize their implementation into the archaeological processes. Françoise Audouze and André Leroi-Gourhan emphasized that, in contrast, French archaeology during these decades was characterized by weakness in theorization and formalization. Insularity is a matter of boundaries, and, therefore, I propose to examine their definitions, starting with attempts at formalization.

Multidimensional analyses (MDA) present some good characteristics for this study. These methods aroused a renewed interest in the field of statistics during the 1960s and 1970s. In France, Jean-Paul Benzecri's *Analyse des données*[3] provided a whole range of complementary and articulated methods for (di)similarity measurement, classification and the graphical synthesis of datasets.[4] In archaeology, these methods appeared to be the most powerful and modern method to deal with data. In some ways they can be considered as characteristic of the New Archaeology, since the first factorial analysis application was performed by Lewis Binford.[5] I aim to focus on this set of methods, but in a country where the New Archaeology has not been so influential: France.

This raises an important underlying question. In his *Domestication of the Savage Mind*,[6] Jack Goody showed how lists and tables were interlaced with the development of the first Middle East writing systems. He considered them as graphical technologies and stressed their cognitive consequences for human evolution. However, Goody's innovative study only took into account data structures with one and two dimensions: lists and tables respectively. What happens if the number of dimensions increases and becomes theoretically unlimited? In a very early reflection concerning applications of principal component analysis in archaeology, the French archaeologist Jean-Claude Gardin emphasized the specificities of archaeological data compared to linguistic or ethnographic data: a large amount of data which are non-linguistic[7] and partially non-numerical. Examining PCA applications, Gardin stated that 'the result is evidently a sharp increase in the combinatorial complexity of the problem, which is likely to exceed the computing power of the human brain'.[8] However, he focused on conceptual issues and did not consider a potential increase in computational power. This increase would happen in the following decades, solving this complexity and providing graphical analysis for n-dimension vector spaces.

The cognitive and social consequences of this new possibility will be our concern here, based on the French prehistoric archaeology case. According to Alberto Cambrosio and Peter Keating, the way a scientific practice acquires an institutionalized form – becomes a discipline – is part of what they call the 'disciplinary stake'.[9] They claim that the notion of 'discipline' involves a concept of identity and boundary control, in order to *preserve them against the danger of heterogeneity of methods and concepts*.[10] What happened in France when MDA, both a new standardized metrology of similarity and a graphical synthesis method, were integrated into the prehistorical archaeology discipline?

To address these questions, I shall combine three kinds of materials: interviews conducted with protagonists in the field; a bibliographical review of the *Bulletin de la Société préhistorique française* (*BSPF*)[11] which published 1,323 papers between 1977 and 2005, including thirty-two presenting an MDA; and lastly a citation network surrounding these thirty-two target papers.

My aim is to show (1) that MDA were carried out by a new kind of actors in this field, who challenged the previous common language shared by prehistorians. This fundamental change was important, considering that (2) language is a fundamental point for the epistemology of archaeology. However, a comparison of MDA applications over time shall make clear that the differences are mostly a generational matter: the transmission processes between them will be addressed (3).

How to Be a Radical Innovator in the Archaeological Field

Innovators' Portraits

Due to its complexity, performing an MDA requires technical skills which were initially not available in the archaeological field. Let's start by sketching the scientific career of some of the main introducers of these analyses in prehistoric archaeology. François Djindjian (1950–) was studying engineering when he applied for the first time to a faculty of archeology. He got his PhD (1981) with a dissertation concerning computer applications in prehistoric archaeology. He conducted a career both as an engineer and firm manager and as a prehistorian, and got a position of associate professor in Paris. His co-author Bruno Bosselin was also an engineer. Georges Sauvet (1941–) is Janus faced. He carried out a career as a chemistry professor in a Parisian university and, separately, developed his own researches about prehistoric rock art. He published numerous papers and acquired a full legitimacy in this field of studies, attending Benzecri's seminars at Jussieu university in Paris. François Djindjian, notably, completed a *maitrise* degree in statistics under his supervision. Some distance from Paris, in Southern France, both Jean Lesage (1923–2004) and Michel Livache (1944–) undertook pioneering MDA applications in prehistoric archaeology. The former was recruited by the CNRS[12] in 1965 as a prehistorian specialized in Meso and South American prehistory. A painter, amateur astronomer and prehistorian, he added statistics to his broad range of interests. Starting in 1971, he decided to undergo his 'recycling' in statistics and began to train himself seriously, making contact with specialists in the domain. Michel Livache was one of the key figures of the *'typologie analytique et structurale'* group of researchers led by Georges Laplace. This movement had its roots in the method developed by Georges Laplace,[13] in which statistics assumed a central role. This group remained in an outside position in the archaeological field. Michel Livache was an elementary school teacher and provided important contributions to this group; he notably learned programming by himself and wrote *'typologie analytique'* computer implementations. All of these researchers were situated in interstitial institutional and cognitive locations, somewhere between engineering, applied mathematics, applied chemistry and archaeology. Only Jean Lesage held a professional position as a prehistorian.

An Aptitude for Mobility

This interstitiality was balanced by a proficiency in mobility, an aptitude for gathering resources from different locations, peoples and instruments. In 1972, Jean Lesage did an internship in the Centre d'analyse documentaire pour l'archéologie

(CADA) founded in Marseille by Jean-Claude Gardin and then led by Mario Borillo. He benefitted from the proximity between the CADA and the Centre de physique théorique (CPT), which made it possible for him to perform analysis on his Mexican lithic tools dataset. This laboratory had at its disposal a terminal connected to the Orsay University's (near Paris) UNIVAC 1108, a powerful computer machine. The combinatorial complexity of MDA required this kind of uncommon machine; to get access to them entailed creating the necessary social relations. On his side, François Djindjian treated his datasets on Saturdays for several years at the Centre Inter-Régional de Calcul Électronique (CIRCE) at Orsay; Michel Livache collaborated with a young statistician, recently hired at the new University of Pau, in order to analyse Chinchon site's lithic industry by MDA.[14] Such examples could easily be multiplied. Let's consider the thirty-two *BSPF*'s papers, related to forty-three authors in total. The collaborations between prehistorians and specialists from other fields (statistics and environmental studies) happened mainly during the 1970s and 1980s. Taking only the authors into account would miss an important point concerning the notion of authorship: the person who performed the MDA could sign or not sign the paper, be mentioned in endnote or not. The period doesn't matter concerning this point.

Dealing with the formal expression of archaeological data pushed archaeologists to develop international scientific collaborations. For instance, François Djindjian was for several years one of the few French prehistorians who participated in the Computer Applications in Archaeology Conference, the most important international conference devoted to this topic. Michel Livache belonged to the *typologie analytique* group, which gathered archaeologists from France, Spain and Italy each summer in Arudy (a village in the Pyrenees Mountains) from 1969 to 1988.

Generic Devices and Legitimacy

Thus, the innovators are characterized by their abilities in both natural and formal languages. This general competence in communication had its materialization in their close relationship with generic devices. Like multidimensional methods, such devices are not especially designed for a particular problem but are theoretically considered to be applicable to a broad range of questions.

We mentioned the requirement of computation centres for the realization of MDA. At the end of the 1970s, personal calculators became affordable, while personal computers appeared in the 1980s. Jean Lesage, Georges Sauvet, Michel Livache and others wrote their own softwares to perform MDA. Their code, written in BASIC or FORTRAN, passed from hand to hand among the archaeologists.

Once they acquired the skills to practice the *Analyse des données* methods, these researchers multiplied the application fields. For instance, in 1999, Bruno

Bosselin and François Djindjian performed a multiple correspondence analysis on sedimentologic, palynologic and palaeontologic data together,[15] even though none of them had any specialized skills in these fields. On his side, Georges Sauvet co-published several papers with a linguist. The software tools they developed were adapted for both archaeological and linguistic datasets. Contrary to Gardin's view mentioned in the introduction, the idea here was that the nature of the objects did not particularly matter: MDA could be performed as long as it was a question of distance measurement between a set of objects. This underlying belief was encouraged by Jean-Paul Benzécri himself about his *analyse des données* methodology and denounced as *benzécrisme*.[16]

An Emphasis on Metrology

This vigorous integration of measurement in the archaeological processes to discriminate artefacts could serve two kinds of strategies. Classifying and defining the most efficient typology were the main concerns of French prehistoric archaeologists from the 1950s to the 1980s. The main typological system was proposed by François Bordes,[17] Denise de Sonneville-Bordes and Jean Perrot[18] and then challenged by Georges Laplace, who contested its intuitive basis in its *typologie analytique*.

The use of MDA was, on the one hand, to defend the validity of previous typology. This was the case in the works of Jacques-Élie Brochier and Michel Livache: a correspondence analysis demonstrated that the C stratigraphical level of Chinchon does not belong to the Magdalenian IV chronological phase but to the Tardigravettian.[19] On the other hand, some authors used this analysis to criticize the old typologies and to build a new one in a more explicit and rational way. André Chollet (a pharmacist), Pierre Boutin and Basavanneppa Tallur (a statistician) claimed that 'rather than using a predetermined system of analysis, such as that of Georges Laplace, it seemed interesting to use the *analyse des données* techniques to try to identify partitions in this set'.[20] They finally concluded that the MDA fitted with the Sonneville-Bordes' typology. For their part, François Djindjian and Bruno Bosselin made a revision of the Laugerie-Haute stratigraphy, an important Paleolithic site. The MDA led them to propose a new typology and a new division of the Solutrean phase.[21]

A Framework for Twentieth-Century Organization of Research

Terry Shinn proposed a framework to characterize an aspect of the twentieth century's 'radical' innovation in scientific activities that he called 'research-technology'.[22] The same features that we previously mentioned are seen as characteristics of this research mode: to be carried out in interstitial areas, to entail generic devices and to have strong concerns with metrology. Most of the sociological approaches (Merton, Ben-David, Abbott) look at science at the discipline level and stress differentiation processes. The 'new' sociology of science has claimed an ethnographic perspective on science and adopted a rather anti-differentiationist perspective on science, between the social and cognitive dimensions. But anti-differentiationism does not imply attention has been paid to integration processes, as Shinn pointed out.

The prehistorians I refer to were involved in two kinds of integration processes. On the one hand, they introduced a new way to present, shape and analyse the archaeological data. They developed a twofold discourse. They proclaimed the power and the necessity of these methods, but they advocated an epistemically careful use of it. In this way, they tacitly included themselves amongst the few able to use them properly. On the other hand, they were able to cross disciplinary boundaries in order to collaborate, especially with foreign researchers. Considering this twofold advantage and control over languages, we are led to further examine the epistemic role of language in prehistoric archaeology.

Multidimensional Patterns and the Problem of Intentionality

Taking into account formalized languages, in addition to the natural ones, is of particular relevance because of the particular status of linguistic facts in archaeology. Archaeologists deal with remains of phenomena which imply a language ability, but the content of this language is definitely lacking in an archaeological site.

This lack can be conceptualized as an incompleteness of the empirical data, compared with those available for the historian or the sociologist. Such an incompleteness can be compensated, depending on the author, by the use of analogy, interpretation or identification of statistical trends. In this perspective, archaeology is mostly a way to get an understanding of the meaning of material remains and to draw a restitution of the past. Let's call this a semantic-oriented archaeology.

In another way, this lack can be seen as an epistemic feature of the archaeological arguments' construction. In this perspective, interpretation is severely criticized and a strict control on archaeological enunciation is requested. The aim of archeology is, here, the study of the transformations of a set of artefacts over time. Let's call this a syntactic-oriented archaeology.

Donald Kelley pointed out the importance of language for the history of science by suggesting that a discipline could be considered as a speech community.[23] In this section, I aim to show the various epistemic roles assigned to the MDA, and the ways the consecutive generations of researchers considered these roles. An MDA is a pattern finder method. The controversial point concerns what can be said about an identified pattern. Thus, we could distinguish the epistemic roles according to whether one aims to deal with the syntactic aspect or the semantic aspect of an archaeological dataset.[24]

Multidimensional Patterns and Syntactic Interest

Most of the oldest papers showing an MDA expressed a very careful attitude regarding the meaning of the results. A pattern identification was considered, so to say, as a result in itself, and the limitations of the methods were underlined, as here, for instance:

> In the present state of research, they [the MDA] do not provide an 'automated sorting machine'; their primary role concerning typology is rather to highlight the structure of a dataset and to allow comparison of related sets later.[25]

In a later paper, Bruno Bosselin and François Djindjian[26] expressed their doubts about any interpretations of their *Analyse des correspondances* in social or psychological terms. The distance measurements and the factorial structure were supposed to be related primarily to an archaeological data structure, and not directly to a psychological or social past reality. This concern was also of central importance for the archaeologists who took part in the *typologie analytique* group, such as Jacques-Élie Brochier and Michel Livache. The emphasis was on issues of methods, and psychological or social interpretations were either considered as a secondary concern or not treated at all.

Multidimensional Patterns and Semantic Interest

A second kind of papers showed an increasing interest in the semantic aspects rather than in the syntactic aspects of the patterns produced by an MDA. In this perspective, a factorial structure was considered, somehow, as a remaining consequence of a past intentionality. This intentionality was interpreted as proceeding either from a collective intentionality or from a personal intentionality.

Georges Sauvet and Suzanne Sauvet[27] performed a multiple correspondence analysis on a prehistoric rock art dataset in order to explain the spatial distributions of the motifs. They claimed that the factorial structures revealed an underlying semiotic system, which is a way to combine both collective and individual levels of action without referring to psychological concepts.

A 1982 study by André Decormeille and Jacques Hinout is of interest here, in that they combined explicitly collective and personal intentionality. Their aim was to discriminate among the various Mesolithic lithic artefacts from the Parisian region.

A correspondence analysis, eventually followed by a cluster analysis, helps to show the population unity or its multiplicity for a given site and to precise the features of the standard tool... a virtual armature whose features equal the mean of those of a given population. It represents the ideal image of the tool that the Mesolithic men suggested themselves to produce.[28]

> It is indisputable that the first factor indicates a significant gap between the north and the south of the Paris basin and the nature of this gap is cultural. ... In conclusion, it is clear that the results obtained demonstrate the power of the method used.[29]

MDA has been used here to fill the gap between the material remains (their proper individual features), an aggregation of individual stone knappers and a holistic entity (here a culture).

Thus, an interest in the semantics of the dataset appeared in some early works. It became predominant in later papers. For instance, Bruno Bosselin aimed to distinguish between 'primary intention productions' and 'derived intention productions'[30] based on the results of a multiple correspondence analysis. The pattern identified was supposed to show the distinction between the tools (what the knapper wanted) and the refuse (what he did not want). Papers published in the 1990s and later presented less frequently the previous methodological concerns: the presentation of the analysis (which algorithm was used, by who?) became less detailed if not lacking, as if this method had become transparent. This phenomenon was described as the black boxing of an instrument.[31] But how to explain the strengthening of interest in the identification of personal intentionality and the relative dismissal of the methodological concerns? Let's take into consideration the reasons why some researchers moved away from MDA.

MDA under Prehistorians' Criticisms

François Djindjian had a close relationship with Jean-Paul Benzécri, whose seminar he followed, and he graduated in statistics under his direction. However, he finally opposed his master on the epistemic role and objectivity of the method. Contrary to Jean-Paul Benzécri, François Djindjian argued for the necessity of a data structuration prior to the performance of a correspondence analysis. In the following years, he developed such a method, dedicated to archaeological applications.

Catherine Perlès also followed Benzécri's seminar and attempted some *Analyse des données* applications on lithic remains. However, she moved away this method, but for a different reason than Djindjan's. According to her, the inferential and non-hierarchical perspective was more of a disadvantage than a benefit. She considered the capacity to hierarchize as a natural competence. However, because of its non-hierarchical basis, the MDA would miss the prehistoric intentional facts. Thus, she turned to the technological study of lithic tools.

In her research on Neolithic stone artefacts, Vanessa Léa scrutinized the concept of 'tool'. According to her, the shape cannot be deemed as a sufficient criterion to identify a tool, considering the morphological variability of the artefacts that she analysed. To solve this difficulty, she turned also to other methods, namely the technology analysis, related to the production of the tool, and the use–wear analysis, related to the use and function of the tool. In doing so, she preserved her research question – the tool – but at the cost of a change of method and a redefinition of the relevant features to analyse.

Results

In archaeology, the identification of intentionality raises a more salient and fundamental problem than in other sciences devoted to anthropic phenomena, in which it is assumed that intentionality could be considered as a non-ambiguous fact. As Arkadiusz Marciniak[32] pointed out, each theoretical view in archaeology is located somewhere between an assumption of reachability and the assumption of unreachability of past intentions based on the study of material remains.

I suggest that MDA appeared as a hope to overcome the gap between prehistoric intentional facts and material remains observed in the archaeological field. By the combination of a huge quantity of data and the quick and automatic extraction of its very structure, it was as if it was possible to reach the inner syntax of a set of archaeological artefacts. This method gave hope of reaching the unreachable: as if the syntax resulting from a formal language would have been an approximation of the structure of a collective or personal past intention.

However, this belief was not broadly accepted. Even if during each decade some exceptions appear, I propose to summarize this historical evolution. I have introduced a distinction between syntactic and semantic orientations concerning the archaeologist's query. I suggest the application of the same distinction concerning the methods used to reach this aim, according to whether they used a natural or an artificial language.[33] Table 3.1 organizes the trends of research in French prehistoric archaeology from the 1950s to the 2000s.

Table 3.1: Trends of research in French prehistoric archaeology

Object/method	Natural language	Artificial language
Syntactic	Empirical typology	*Typologie analytique*
Semantic	Technology	Synthetic lineage

'Empirical typology' includes works by both François Bordes and André Leroi-Gourhan. Leroi-Gourhan has clearly expressed a syntactic interest, for instance:

> Techniques are at the same time gestures and tools, organized in sequences by a true syntax which gives to the operational series both their stability and their flexibility.[34]

'*Typologie analytique*' and 'Synthetic lineage' were most related to the integration of MDA into the prehistoric field. A syntactic-oriented method, such as MDA, does not imply an aim of knowledge oriented toward the syntactic aspect of the data. Finally, the latest orientation, namely technology studies, turned again toward a method relying on natural language. Let's assume that this periodization is relevant and that we have such 'generational paradigms', following Andrew Abbott's words.[35] How can we explain the genetic differences between each generation? What kinds of transmission processes bound or divided them?

Intergenerational Boundaries and the Transmission of Methods

Citation analysis[36] is a way to study transmission processes at a larger scale of analysis. A citation network which surrounds the thirty-two papers published in the *BSPF*[37] from 1977 to 2005 has been obtained from Google Scholar.[38] It counts 2,589 edges (citations) and 2,112 vertices (papers). Assuming that to cite a paper reveals a cognitive interest for it, where are the MDA-based papers located in the archaeological bibliographic field? What kinds of relationships did they develop over time?

To start with, a basic observation can be made regarding the relationships among the thirty-two target papers. Without consideration for the chronology, fifteen papers among the thirty-two have a tie with at least one other. From a bibliography standpoint, around half of the multidimensional applications were done without referring to papers which had a similar interest. Among the connected papers, three groups can be distinguished: a cluster of pioneering applications (including works by François Djindjian and Georges Sauvet), a cluster of papers concerning the Mesolithic period (including works by André Decormeille and Jacques Hinout, who we mentioned) and a cluster constituted only by the self-citations between papers co-authored by Bruno Bosselin and François Djindjian. This network is faintly connected. We shall conclude that the *BSPF* in itself was not an especially favorable publication space for the development of MDA. There was a need for editorial spaces devoted to computer applications in archaeology: small journals were created for this concern, but remained confidential (for example *Archéologues et ordinateurs*). More generally, this lack of editorial space helps to explain the low diffusion of MDA in French archaeology.

If the target papers themselves are not especially bound together, what about the references related to the MDA bibliography? Considering that the pioneering archaeological applications of such analysis cited publications which specialized in statistics, what can be said about the transmission of this indexation? To answer this, I built for each year a network including: (1) the ties created during the year considered and directed toward a target paper; and (2) the methodology references cited by the target and, if it is the case, also by the level one papers. Did the papers that cited a target paper also cite its methodology references? Formally, this can be answered by a measurement of the network transitivity for each year. Figure 3.1 shows clearly a dual phenomenon of disconnection of the archaeological literature from the methodology references.

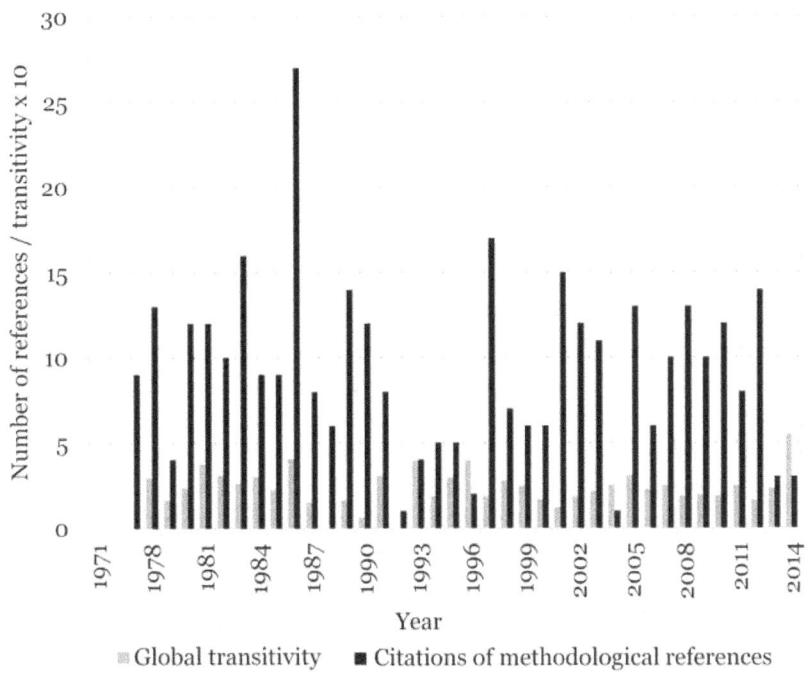

 Global transitivity ■ Citations of methodological references

Figure 3.1: Citations of methodological references (black) and global transitivity (grey)

Starting from around 1995, the number of methodology references decreased and the transitivity score fell and stayed at zero. This result strengthens our previous observation concerning how the authors of the MDA receive less, or no, mention over time. This can be seen as evidence of integration of the method into the archaeological field. However, this integration has to be nuanced: the transitivity score, even before 1995, has never been very high: either the prehistorians have become so qualified with regard to MDA that they were able to disregard the relative literature, or they simply neglected its methodology.

This is mostly a matter of education, which is the other main vector of transmission in scientific activities;[39] however, it cannot be reached by the sole citation analysis. There were two principal ways to gain the competency to perform an MDA: a quite informal teaching from one person to one, let us call it dyadic transmission; and the institutionalized frame of transmission, for example lectures or summer schools.

Dyadic forms of transmission mainly appear for contingent reasons due to the personal network of a researcher. For instance, in the 1990s, Jean Lesage welcomed into his own home and personally helped a young prehistorian throughout his thesis. The second form appeared often in quite marginal academic spaces: the *séminaire de typologie analytique*, previously mentioned; the 'European Summer School' organized by François Djindjian in Valbonne and Montpellier (1981 and 1983) or the seminar he led at the *École Normale Supérieure*, starting in 1985.

Conclusion

In this paper I have explored three lines of evidence to shed light upon the integration and reception of MDA in French prehistoric archaeology. I reintegrated the history of this set of methods into the main trends of development in this field. The role of a particular type of protagonist, the combination of these analyses with the fundamental issue of archaeology – through the case of intentionality – and, lastly, the bibliographic structure of their reception were considered. The promotion of MDA was part of a broader call for a rationalization, a standardization and an explication of the reasoning of French archaeology. What changes did this integration induce into this field?

Since Kuhn's *Structure of Scientific Revolutions*, a classical explanation of scientific change is a paradigm shift which occurs when too many anomalies are raised in the normal science: consequently a differentiation process takes place. This explanation may be appropriate when it comes to physics but it might not be relevant for prehistoric archaeology. An 'anomaly' set, which would have been inconsistent with the morpho-typological framework, never emerged. I suggest here that the integration of the *Analyse des données* and then the lack of interest about it should be explained by some more contingent reasons, related both to the structural and the biographic dimensions of scientific activities. Terry Shinn's 'research-technology' framework for twentieth century science seems to be relevant to explain the intervention of the innovators who introduced multidimensional analysis, thanks to their abilities to move between several interstitial areas and to reach the prehistoric archeology field. Rather than anomalies, we should speak here of a cumulative replacement of questions, what Andrew Abbott called 'object inflation' in the field of sociology.[40] He proposed a dynamic model in which discipline development is moved by a fractal process of lineage differentiation at each generation. In these cycles, steps of division and conflict among lineages can be followed by an ingestion process: the victorious lineage integrates the questions and methods of the defeated lineage. Such a model can now be proposed, based on what we have learned from the MDA applications case in prehistoric archaeology (Figure 3.2).

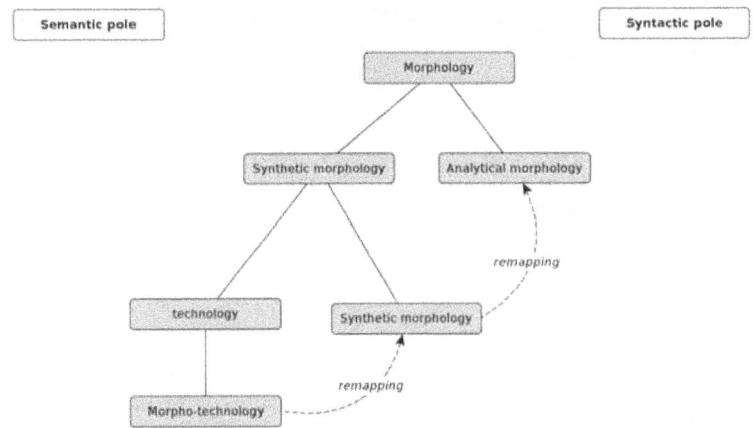

Figure 3.2: A model for the development of prehistoric archaeology

In the decades following the Second World War, the morphological approach was the main trend for lithic artefacts studies. From the end of the 1960s on, proponents of an analytical typology opposed the natural language-based and intuitive methods of the previous generation. However, the first lineage resisted and became hegemonic. Some aspects of the analytical branch were remapped and integrated: formal methods were considered as a potential way to capture collective intentional facts. This dominance was then challenged by a flourishing lineage, which promoted a technological analysis from the mid-1970s on. The newcomers criticized the morphological approach for being limited to aggregated phenomena (culture, society) and for missing the personal dimension of prehistoric life. Rooted in Leroi-Gourhan's palethnology, the technological approach aimed to reach the culture and, above all, the 'man behind the tool'. It became in turn the dominant trend in the mid-1980s, and after a while integrated some features of morphological analysis. A proper lithic study was then supposed to combine both of these analyses. In this paper, little space was given to another important lineage, since it was not as related to the formal methods, namely the functional (use–wear) analysis and the recently flourishing archaeometry.[41]

Does it mean that nothing changed following the integration of MDA? Writing the history in which he took part, François Djindjian did not hesitate to label the 1970s as a 'Golden Age';[42] for his part, British archaeologist and statistician Mike Baxter claimed that most of what has been done since this decade has never been of such radical novelty;[43] Jacques-Élie Brochier, a former *typologie analytique* promoter, criticized his colleagues' lack of rigor in their data analysis treatments.[44] Obviously, summing up the consequences and the legacy of an intervention into a scientific field is also a matter of self and generational legitimation. Indeed, Abbott's view on scientific evolution challenged the very notion of cumulativity regarding social science knowledge; a genuine scientific breakthrough is likely to be found elsewhere, not in the claims for innovation that occur in each generation. As shown by Wiktor Stoczkowski, anthropogenesis narratives dating from the sixteenth to the twentieth centuries can be reduced to a combination of a few elements.[45] Facing this long-standing persistence, it is doubtful that any standardization attempt could ever radically change the deep (social) motives of prehistoric studies.

Acknowledgements

I would like to thank the members of the *Centre d'étude des rationalités et des savoirs* (Toulouse, France) and the members of the *Centre d'Història de la Ciència* (Barcelona) for their comments on earlier versions of the paper. I would also acknowledge Lara Huber and Oliver Schlaudt for their help. A particular thank you is offered to Aurélien Tafani and Kathryn Ticehurst for their proof-readings.

4 THE DOUBLE INTERPRETATION OF THE EQUATIONS OF PHYSICS AND THE QUEST FOR COMMON MEANINGS

Nadine de Courtenay

The use of genuine mathematical equations to express physical relations is a surprisingly recent phenomenon. Although expressions that look very much like the familiar equations of classical physics we have been taught in school were used in eighteenth-century rational mechanics, closer inspection reveals that these expressions are really 'simple proportionalit(ies) indicated in an abbreviated manner'.[1] The rationale for our current use of textbook equations can be traced to the 1950s, but it is unlikely that many of the equations' contemporary users will have considered the scaffolding that enables us to handle these equations as we do. In the following, I will try to give a sense of how physical relationships have come to be represented in mathematical form, and show that the reason why we can securely embark on our current manipulations, even though we largely ignore their theoretical underpinning, is because these manipulations are supported by a hidden structure that permeates the formalism we have learned to use.

When we deal with textbook equations we casually move between two distinct interpretations of the mathematical expressions we are handling. This 'double interpretation' of the equations of physics was first introduced by Maxwell in an article of the *Encyclopedia Britannica* entitled 'Dimensions':

> There are two methods of interpreting the equations relating to geometry and other concrete sciences.
>
> We may regard the symbols which occur in the equation as of themselves denoting lines, masses, time &c.; or we may consider each symbol as denoting only the numeral value of the corresponding quantity, the concrete unit to which it is referred being tacitly understood.[2]

I will show that these two interpretations derive from two distinct ways of putting the experimental physical relationships of the classical sciences into the form of mathematical equations: the mathematization of the Baconian sciences in the nineteenth century prompted, first, an interpretation of the equations of physics in terms of *measure equations* and, secondly, an interpretation in terms of *quantity equations*. Both mathematization schemes heavily depend on units although they rely on quite a different understanding of the latter's nature and role. Units will thus turn out to be important not only for measurement and for communicating our knowledge of quantities; they also underlie the way in which we represent physical relationships mathematically and, hence, have a bearing on the problem of the applicability of mathematics to physics. Acknowledging the role of units will make us realize that if we have been oblivious to the way in which the two paths of mathematization mentioned above have come together in the double interpretation of the equations of physics, then we have also overlooked the human thread running through the philosophical analysis of our most basic scientific tools.

The Numerical Interpretation of the Equations of Physics

Fourier and the Mathematization of the Science of Heat

It is by enlarging on the Cartesian tradition underlying rational mechanics that Fourier succeeded in submitting the phenomena of heat conduction to mathematical treatment. The framework of the mathematization process he developed in his *Théorie analytique de la chaleur* (1822) still rested on the fundamental idea that quantities of different kinds could enter into quantitative relations with one another, be combined and treated mathematically, only by means of the ratios they could form with quantities of the same kind. But Fourier parted with the eighteenth-century tradition by replacing the undetermined quantities at the denominator of these ratios, that were previously ascribed on a case-by-case basis, with specified, particular quantities.[3] Innocuous as it may seem, this move laid aside the abstract character of rational mechanics by turning the hitherto undetermined ratios between comparable quantities into numerical functions that everyone could assign in the same way. The mathematical representation of physical relations between quantities thus became widely regarded as functional relationships between *measures* of quantities, which, as numbers, were, of course, straightforwardly amenable to mathematical treatment.

The selection of a preferred set of units is of paramount importance in the transition from the abbreviated ratio equations, still lurking behind the abstract equations of rational mechanics, to a purely numerical understanding of the equations of physics as genuine mathematical equations between measures. Indeed, the values of the numerical variables or functions standing for quantity measures can be assigned, and the numerical constants appearing in the functional equations experimentally determined in a definite way, only if: (i) the quantities occurring at the denominator of the corresponding ratios become particularized thanks to explicit definitions; and if, correlatively (ii) the instructions needed to *materialize* these particular quantities as standards are clearly provided. These definitions and instructions are indispensable if the mathematical equations of physics, interpreted as equations between measures, are to be given the *same meaning* by different individuals working apart in different settings. In particular, the proportionality factors, which appear when abbreviated ratio equations are transformed into genuine equations, take the same specific numerical values for everyone. Prior to the nineteenth century it made little sense to gather and publish such factors in numerical tables of 'physical constants'.[4] In this regard, the transition from ratio to assignable functional equations involves a collective organization whose impact on the inner working of science, industry and trade is seldom acknowledged. By ensuring the communication and sharing of numerical equations and results (especially of the numerical constants involved in the equations), this organization not only enables one to use the findings obtained by others and combine them with one's own; it also promotes a gain in testability and inferential power by setting a common stage on which a collective critique of the results obtained can meaningfully take place.

In carefully spelling out the units and scales to be used in his *Théorie analytique de la chaleur* before proceeding to establish the equations of heat conduction, Fourier was, no doubt, emulating the course Euler had began to follow in mechanics. But Fourier's move was actually taken as a pattern because it blended with the economic and practical concerns of the industrial revolution and the initiation of the metric reform.[5] This reflects how much the conceptual framework governing the transition to the numerical interpretation of the equations of physics, introduced as a means to achieve the mathematical representationof the Baconian sciences, is deeply entwined with the social, economical and institutional demands for the communicability and coordination of scientific results and activities.

The philosophical ramifications of the numerical conception of the equations of physics were spelled out by the positivists at the end of the nineteenth and twentieth centuries. But, as we will see, their empiricist stance led them to reinforce the role of units and, hence, to endorse a conventionalism that cast aside social issues.

Measurement, Conventions and the Reconstruction of Physical Concepts

Positivism gave a new twist to the numerical conception of the equations of physics. The reference to ratios was discarded to stress even further the conceptual role of units, standards and measurement methods.

According to the positivists, the building of standards and the empirical procedures used to assign numbers to the variables involved in the functional equations of physics provided a way to *reconstruct* the physical concepts corresponding to the variables in the equations. Following Duhem, the reconstruction was presented as a *translation*: the choice of a specific empirical procedure served as a dictionary that provided a translation ('version') of the qualitative language used to denote properties encountered through experience (such as hotness) to symbols taking numerical values (temperature, denoted by the letter T which should be indexed by the procedure). The empirical procedures underlying the rules of translation had to exhibit certain structural properties if the mathematical manipulations in the realm of symbols were to translate back ('theme') into adequate empirical statements and thus realize a fruitful match between qualitative relations and mathematical operations so as to secure the applicability of mathematics to experience.[6]

This reconstruction based on procedures of numerical assignment epitomizes the verificationist conception of meaning according to which the meaning of a concept is tantamount to a rule enabling the determination, on the basis of observation, of when the concept applies (say, having a temperature of 25°C). However, this account of meaning propounds a purely numerical understanding of the concepts and equations of physics that dispenses with the assumption of pre-existing quantities. Moreover, it leads to the construction of as many quantitative concepts as there are empirical procedures available for one and the same qualitative attribute. This diversity can only be reduced by conventionally selecting, for each case, a particular construction method, that is a particular standard or concrete scale and a particular procedure.[7]

The idea that the construction of physical concepts is intrinsically interwoven with the conventional choice of standards and concrete scales became even more compelling to the logical positivists reflecting on Einstein's relativity theories. The judgement that two rods, locally observed to be congruent, remained 'equal in length' when moved apart in different locations seemed to require an additional preliminary rule fixing the class of material bodies that would be considered as rigid. Only a 'coordinative definition' connecting the concept of a rigid rod taken as unit with particular concrete objects regarded as materializing the unit would enable each individual to define length and turn statements about length into characterizations of objective states of affairs.

The positivist reconstruction of the concepts of physics imparts a *constitutive role* to the coordinative definitions, and other choices of standards and scale: these coordinative definitions are indeed, in a way, a priori since they are not derived from experience and serve to determine the physical concepts and empirical laws. However, if they meet the needs of the verificationist, individualistic criterion of meaning they also fix the expression of the empirical laws and make it difficult to account for the collective nature of scientific activity. To be sure, a surrogate for the lack of necessity and universality of the coordinative definitions and choices of standards in general can only be retrieved in the form of a consensus among independent individuals willing to take into account the choices of others in order to develop a shared language of physics. Yet, it is very difficult to see how such a conjunction of already 'objective' individual representations and decisions can be achieved and result in a social, coherent activity, open to evolution.

Indeed, one should acknowledge with Descombes that there is more to social practices than individual activities taking account of one another: a distinction should be made between *shared meanings* as *intersubjective* meanings and as *common* meanings.[8] As a matter of fact, one's measurements have to coordinate with *other measurement activities one knows nothing about*, including *possible* activities that might never actualize. Only something over and above intersubjectivity, like institutions, can support this kind of interrelated activity. This is apparent in the scientific and technical activity of subjects who have to cooperate in a complex society displaying high division of labour where they have to assume different complementary roles and purposes.

In this respect, the external, historical dimension deliberately left aside by the reconstruction process sketched above proves to be essential: the activities that the rules are supposed to recreate have been exercised for trading and administrative purposes from the dawn of civilization; settling on one choice among different other possible conventions would involve *regulative* rather than constitutive rules. The need is indeed to regulate and resolve eventual conflicts arising between different practices and different expressions of empirical laws already in use, based on the different sets of units and standards that individuals, or rather communities, have actually been led to endorse as a result of complex historical processes.

This regulative function has to deal with both the coordination of symbols with the world and another type of 'coordination': the coordination of the users of these symbols involved in *cooperation*. In addition to the structural conditions underlying the translatability between a mathematical language and the qualitative language articulated to phenomena ($Transl_1$), one also has to attend to the conditions that have to be met to ensure the translatability between *different possible mathematical representations* ($Transl_2$) – such a translatability will turn out to be constrained by *invariance conditions*. Taking simultaneously into account the relations of subjects to the world and of subjects with one another introduces, along with the referential conception of meaning entailed by the positivistic framework, the dimension of meaning as *sense*. The appropriate epistemology is therefore not the classical individualistic epistemology but social epistemology. This is not to say that knowledge is a matter of consensus – such an idea stems from a classical epistemology trying to patch a social surrogate for universality out of an individualistic conception of meaning. 'Social', here, does not point to what is simply public or shared, but as Descombes and Longino[9] suggest in different ways, to a complex *interrelatedness*. Common meanings turn out not to depend on the selection of a specific set of standards adopted by all, through which symbols can be directly linked to the world thanks to individually controllable verification criteria (i.e., standardization). Instead, they rely on the practical requirement (bound to evolve with technical and social degrees of sophistication) that subjects adopting different standards, and therefore different systems of representation, can cooperate effectively through instrumented interactions even if they have to mutually translate their findings in one another's systems. Such a framework conjoins communicability and translatability to comparability and cooperation. It invites a conception of verification in terms of collective (rather than mutual) coordination and sets the stage for the pursuit of extensive critical activity, and for the refinement, displacement and creation of meanings. A way to handle at once the plurality of individuals that are engaged in scientific and technical activities and the connection of these activities with the world is to tackle the problem of changes of units.

The Equations of Physics as Equations between Quantities

Towards a New Perspective: The Problem of the Change of Units

The variety of units, standards and scales isn't only a consequence of historical contingencies; different practical aims and areas of physics tend to select different scales and procedures adapted to their purposes – astronomy and surveying can easily illustrate this point. Division of work in a complex society has therefore deep epistemological ramifications: it prompts thorough, separate investigations of the different aspects under which physical systems and phenomena can be considered and, at the same time, requires the coherent recombination of these aspects in the realization of common goals.

The project of the metric reform and the practical interests of the nineteenth century that paved the way to Fourier's numerical conception of the equations of mathematical physics, were both deeply connected with the problems that arose from the use of different sets of units. Fourier paid due attention to what happened to his functional equations when the units used to make the numerical assignments were changed. Relations of proportions between quantities are by construction indifferent to changes of units. But when physical relations are expressed in terms of measure equations, such an indifference is completely lost: the choice of a specific set of units does not only affect the values that will be given to the variables of the functional equation; it shows up in the numerical factors which enter into the picture when one shifts from equations of proportions to genuine functional equations. Indeed, numerical factors such as k in equation (4.1):

$$q = k. p^a. r^b ... s^c \quad (4.1),$$

where $q, p, r ... s$ stand for the *measures* of physical quantities, are *settled by experiment* once the units employed to assign the different measures involved are fixed; and these factors change as soon as one of the units is changed.

However the numerical equations, which express physical relations in terms of a chosen set of units (involving $Transl_1$), should remain valid when they are translated into another possible set of units ($Transl_2$). Fourier's argument was that if mathematical equations express physical relations they should not depend on the contingent choices of units we make to assign the values of the quantities.[10] They should instead *allow changes of units*, that is, *remain valid* for all the choices that can be made to perform the numerical assignments.

To deal with this issue, Fourier stated the general rules showing how the numerical factors appearing in his equations of heat conduction should be modified so that the equations remained valid when the size of the initial units and scales were changed by certain factors.[11] Fourier's seminal study, later, develops into what is now called 'dimensional analysis'. Using today's notation for the purpose of illustration, Fourier's analysis applied to equation (4.2):

$$f = 6.674 \ 10^{-8} \ m \ m' / l^2 \quad (4.2)$$

states that the constant *6.674 10^{-8}* determined by experiment when f, m, m' and l denote respectively the measures of the force of attraction, of the masses of two bodies attracted towards each other, and of the distance separating them, in CGS (centimetre, gram, second) units (in which f is measured in dynes) should be changed, or translated ($Transl_2$), according to the formula $L^3 M^{-1} T^{-2}$; that is, if the units of length, mass and time are multiplied respectively by λ, μ, ν, then, the experimental constant should be divided by $\lambda^3 \mu^{-1} \nu^{-2}$. The constant in equation (4.2) therefore becomes 6.674 10^{-11} in MKS (metre, kilogram, second) units – where f is then measured in newtons. It should be noted that the *same physical relation* is written as (4.2) in the CGS system and written *differently* as (4.2') in the MKS system:

$$f = 6.674 \ 10^{-11} \ m \ m' / l^2 \quad (4.2')$$

It is interesting to note that it would be possible to regain the independence with respect to units displayed by the relations of proportions, while retaining the advantages of measure equations, if the mathematical equations of physics could be understood as equations between the *quantities themselves* instead of between their measures. The independence could then indeed be deliberately reintroduced *in the equations* through a condition pertaining to the mathematical *form* of the equation: the form of the equation should remain the same, be *invariant*, under (certain) changes of units. In this case, the numerical factors in the equations would be considered to be the values of certain secondary quantities[12] characterizing, either: (i) a particular system or substance, as in equation (4.3):

$$M = \rho V \quad (4.3),$$

where the *quantities* M and V are respectively the mass and volume of samples of a substance [X], and ρ is a *dimensioned* constant denoting a secondary quantity – the volumetric mass of [X]; or (ii) a single valued quantity, when the factor remains the same for all systems, as in equation (4.2) – it is then a dimensioned 'fundamental constant', here the gravitational constant G. If the symbols F, M, M' and L are understood to denote the quantities themselves, then equations (4.2) and (4.2') would be rewritten as:

$$F = G M M' / L^2 \quad (4.4)$$

whose mathematical form remains the same regardless of the units employed. In equations (4.3) and (4.4), it is only the *numerical values* of the secondary quantities ρ and G that change (according to Fourier's rules of conversion), provided the units of these secondary quantities are properly defined in terms of the units of the primary quantities involved (length, time, etc.).

But, in Fourier's time, the conceptual, technical and social conditions for endorsing such a conception were far from united: no scheme was yet available that could challenge the numerical interpretation of the equations of physics because there was no way to legitimate the mathematical manipulation of quantities themselves, especially the mathematical combination of quantities of different kinds.

Towards Equations of Physics as Equations between *Quantities*: Maxwell's Method of Mathematization

The stimulus to interpret the equations of physics as equations between quantities came from another endeavour to extend the realm of mathematization into the Baconian sciences. When he set out to establish the equations of electromagnetism, Maxwell proceeded, like Fourier, from experiment, not from theoretical hypotheses; but the kind of experiment involved in his case was not measurement. What Maxwell was aiming to work out with his early method of analogies was a mode of mathematical expression that could capture and extend the spatial, *qualitative* aspects of the experimental relations between electric and magnetic quantities that Faraday had established, as opposed to quantitative features of experiment. In the later, more abstract and mature expositions of his fully grown theory, in the 1870s, he substantiated his avoidance of Fourier's line of attack by criticizing the latter's roundabout, Cartesian way of submitting geometrical and physical quantities to algebraic manipulations.

According to Maxwell, such a line of attack amounted to introducing, from the start, a system of coordinates in which units were used to represent quantities by symbols denoting numbers. All quantities thus came to be treated in the same way, exclusively from the point of view of their size, without attending to their quality and possible direction. Against this approach which made number 'rule the whole world of quantity',[13] Maxwell insisted that physical reasoning required a conception of mathematics that did not reduce to 'calculation', and which enabled to operate upon symbols that could lay hold at once of the quantities themselves. He recognized the seeds of such a view, first, in the development of symbolic algebra in England[14] and, later, in Hamilton's theory of Quaternions. Both of these threads eventually contributed to challenge the idea that mathematics was the science of quantity (as professed by tradition), or a science grounded on the concept of number alone (as claimed by the 'arithmetization movement' of the nineteenth century), and inclined to regard mathematics as a science dealing with relations and structure.

Maxwell successively made use of these two threads in order to avail himself of quantity equations. Thanks to symbolic algebra, a better appreciation of the way mathematical reasoning depended on relations revealed how quantities themselves, and not just their ratios, could be treated algebraically. It was realized that the ability to carry out mathematical deductions and operations didn't rest on the meaning of the symbols manipulated – that is, on the nature of the entities involved – but had to do, instead, with the properties of the relations or laws of combination that existed between these entities. As reminded by Lodge, another promoter of quantity calculus, what defined the arithmetical operations were indeed formal laws of operation: associativity, commutativity, distributivity.[15] In this respect, the applicability of the mathematical operations of arithmetics in physics was not elucidated anymore by the construction of a language summoning translation rules based on choices of concrete scales leading into the realm of numbers. Mathematical operations could *directly* apply to symbols denoting quantities by virtue of a structural similarity between the relational system of numbers (associated with equality, order and addition) and the instances of a given kind of physical property (associated with physical relations of comparisons and concatenation obeying to the same formal laws as the arithmetical relations). This insight underlies the clarification of the concept of measurability Helmholtz presented in his famous article entitled 'Counting and Measuring', which offered, by the same token, a justification of the use of mathematics in physics. The insight is also the gist of the method of analogies through which, Maxwell, following in the footsteps of Thomson, got hold of the equations for electromagnetism.

Hamilton's theory suggested how combinations of quantities of different kinds could be handled to form genuine quantity equations. It implied the building of a genuine geometrical calculus that stood in contrast to the spirit of Cartesian analytical geometry. This calculus drew the attention away from size, that from the start led to extrinsic considerations of coordinates and numbers, and focused instead on *variations of size*, that is, *direction*, which promoted an intrinsic treatment of quantities. The British reception of continental differential calculus provided a means to recast the Newtonian conception of geometry, according to which geometrical figures are not given but *generated* by movement, in the terms of a calculus of operations where geometrical (but possibly also physical) quantities could be *produced* by the application of algebraic and differential operations performed on a set of independent base 'elements', or 'units'.[16] These units are of a different nature and play quite a different role to that ascribed to units in the Cartesian scheme: they do not serve to reduce already given quantities to lines and numbers; their function is, on the contrary, to *generate* hierarchically organized quantities of different kinds.[17] They are thus tantamount to base vectors out of which geometrical and physical quantities can unfold once the algebra specifying their rules of combination is defined. The construction of such an algebra, Maxwell anticipated, would provide a method of 'conceiving dynamical quantities' that would accomplish in physics what Hamilton's calculus of Quaternions had achieved in geometry.[18]

It is in this context that Maxwell and Lodge appended to the numerical interpretation of the equations of physics, adjusted to deal with experimental issues, an intrinsic interpretation in terms of quantity equations better adapted to theoretical purposes. However, the two schemes did not fit together neatly. As we will see, scientific developments were needed to clarify the status of the quantity calculus and show how quantity and measure equations were articulated to one another; whereas institutional developments were required to actually bridge the two interpretations through the construction and adoption of a coherent system of units that dovetailed the theoretical and practical uses together.

The Double Interpretation of the Equations of Physics

The 'Classical', Experimental versus the 'Modern', Theoretical View and the Distinction between Quantities and Magnitudes

The clarifications just mentioned were tardy; they were an outcome of a decision made at the ninth Conférence générale des poids et mesures (CGPM) of

1948 to add to the three mechanical base units of the MKS system a fourth base unit, the unit of electrical current, the ampere, leading to the four-dimensional MKSA system.[19] The question of how to understand the mathematical expression of physical relations was stirred up by heated debates concerning the numerical coefficients that should be introduced in the equations of electromagnetism when they were expressed in the new system. The technical debate illuminated the possibility that physical relationships could be put into the form of mathematical equations in two different ways.[20]

The scientists involved in the debate identified what they called a 'classical' view which converges with the interpretation of equations as measure equations presented above. According to this view, the equations of physics expressed 'the way in which the measures of certain dependent variable physical quantities depend on the measures of other independently controlled physical quantities'.[21] The other outlook was labelled the 'modern' point of view and is kindred to the interpretation of the equations in terms of quantity equations. Indeed, it was pressed home that, owing to the advancement of group theory and algebra as a whole in the twentieth century, Maxwell and Lodge's insights had been elaborated into a fully fledged quantity calculus by Wallot, Landolt, Fleischmann and others. However, whilst putting the quantity calculus on firm ground, these advancements promoted an understanding of the equations of physics which was different from the former understanding in terms of quantity equations in one important respect.

The expression of physical relationships by mathematical equations was now considered to be the result of the construction of a *'mathematical model'*. The elements of this model, the denotation of the symbols in the equations, did not stand for *concrete quantities* any more, but *abstract*, mathematical concepts, only indirectly related to the physical concrete quantities manipulated in experience.[22] These abstract entities, or *models* of the concrete quantities, were introduced postulationally and given *by definition* the property of being subject to additive and multiplicative operations. The long-standing difficulty that it made no sense to perform multiplications or divisions, or other more complex operations on symbols denoting physical attributes was thus completely circumvented. The combination of the *abstract* (symbolic) quantities of different kinds involved in the multiplicative algebra has its result *assigned* so as to comply with group structure and reflect the relations established between the *physical properties bearing the same name in experience*: '(t)hus we do not ask "*How* does one multiply mass by acceleration"? We *define* the multiplication by the statement "The product of mass and acceleration is force"'.[23] This mathematical reconstruction of empirical relations creates a new algebra in which abstract quantities can be combined to obtain new relationships and introduce new abstract quantities in the mathematical model, the algebra being in excess with respect to the network of experimental physical relations that has inspired it.

It was realized that a division of views regarding the mathematical presentation of physics had constantly prevailed among physicists and fostered countless confusions that had gone largely unnoticed because both parties used the same terms to denote very different entities. Indeed, as König and others pointed out,[24] the concept of quantity merged together, under the same name and symbol, two distinct meanings which had to be carefully distinguished: on the one hand, that of a *concrete* quantity tackled in experiment and, on the other, that of an *abstract* quantity tackled in our mathematical models of experiment. Russell had already urged the separation of the two notions and proposed to retain the term 'quantity' to denote *concrete* quantities encountered in experience, and to call '*magnitudes*' the *abstract quantities* involved in the mathematical expression of physical relationships.[25] The least that can be said is that neither the recommendation of the metrologists of the 1950s nor the proposal of Russell have been retained by the scientific and philosophical communities. It must be admitted that usage is often too entrenched to give way to a more rigorous terminology – in the following, I will use the term 'magnitude' where it is appropriate, but I will continue to use the expression 'quantity calculus'.

Bridging Magnitude and Measure Equations: The Construction of a Coherent System of Units

The international committees in charge eventually settled on a solution which straightened out the issue[26] but, at the same time, reinforced ambiguities. Quantity calculus was designed as the proper tool to handle the mathematical equations of physics *which were to be regarded as expressions relating magnitudes*. This settled, an apparently seamless passage was established from the magnitude equations used in theoretical reasoning, which were *independent* of measurement units and of empirical procedures of measurement, to the corresponding measure equations, indispensable to test the mathematical models, and to all practical purposes in general. The smooth passage between the two interpretations was achieved by the construction of a *coherent* system of units which ensured that the equations had *exactly the same expression under both interpretations*. The coherent system of units, which was to be disseminated worldwide by the metrological international organizations, reconciled the two formerly conflicting points of view within a framework that warranted, at last, the double interpretation of the equations of physics propounded by Maxwell in his *Encyclopedia Britannica* article. The cogency of this double interpretation is now taken for granted by every student of physics when solving a problem and sliding from the algebraical solution to its numerical application. The institutional and technological scaffolding which has supported this symbolic practice since the implementation of the MKSA system conceals the obstacles that had to be overcome through a combination of scientific and social occurrences.

The principles guiding the construction of a coherent system of units are roughly the following.[27] Trusting the accepted physical relationships that can be stated in relations of proportions, one chooses among them the relations that will be taken to be invariant by a change of units; the numerical coefficients in these equations are set equal to one, for example $F = Ma$. These equations determine the *structure* of the system, and impose certain relations between the units of the magnitudes involved, here, between F, M and a (note that one starts here with a *structure*, not conventions on units). One then chooses among these magnitudes those that will stand for the base magnitudes of the system, out of which all the others will be generated. This determines the *type* of the system. All the physical relations that do not belong to the structure equations (and are not consequences of them) will then be *made* invariant by changes of units by introducing certain *dimensioned* constants, in other words secondary (or derived) magnitudes that can either correspond to system dependent characteristics (like the volumetric mass ρ) or to system independent, single valued, characteristics, the so-called fundamental constants (like G). Only in the end do the units of the base magnitudes, called *base units*, get selected. The units of the derived magnitudes are then fixed as derived units expressed in terms of the base units through the structure, or the other equations, without introducing any numerical coefficient (i.e., by setting all the coefficients equal to one). It is this last step that is all important: it establishes the bridge between the two interpretations and at the same time conceals the gap between them because it ensures that the equations interpreted in terms of magnitudes and in terms of measures have exactly the same expression. Indeed, the (magnitude) equation for the kinetic energy, $E = \frac{1}{2} MV$, is the same in any system of units having the same structure equations as the International System of Units (SI):

$$E = \{E\} [E] = 1/2 \{m\} [m] \{v\}^2 [v]^2 = 1/2 \{m\} \{v\}^2 [m] [v]^2 \quad (4.5),$$

where $\{Q\}$ and $[Q]$ denote respectively the measure (numerical value) and the unit of the magnitude Q. Indeed, in a coherent system of units, the derived unit of the energy will be $[E] = [m] [v]^2$, and the measure equation has therefore the same form and coefficient as the magnitude equation: $E = 1/2 \{m\} \{v\}^2$.

Working out Common Meanings – Objectivity, Translatability and Uncertainty

Although they appear seamlessly connected, the two interpretations do not have the same status and, actually, do not fit together. As a matter of fact, the gap that opens between them proves to be crucial for the ability of our mathematical representations of physical phenomena to fulfil their function. For one thing, the mathematical equations of physics, in their theoretical use, are primarily to be understood as equations between magnitudes, not between measures or quantities. With this in mind, it is time to take stock of the profound difference between the positivist 'classical' approach and the 'modern' one.

The reconstruction of magnitudes and magnitude equations on the basis of accepted empirical relations is not spelled out in terms of empirical procedures involving coordinative definitions and choices of standards; it proceeds from the *inception of a relational system*, a *structure* from which the definitions of the physical magnitudes that do not belong to the basis get derived. This draws attention to the fact that the physical relations between properties of different kinds may be represented within *different systems*, that is, according to different *choices of structural equations and base magnitudes*. This is not a mere restatement of the conventionalist claim that there are many distinct, and indifferent ways (languages) to express physical phenomena mathematically. The global standpoint taken by the modern approach puts the matter in an entirely new perspective by requiring that the different possible representations should, in effect, be translatable into one another.

In this respect, communicability and agreement are not based on the adoption of one particular set of units, but on the investigation of what is translatable between different permissible choices of systems of mathematical representations. The modern viewpoint is thus deeply connected with the shift in the conception of meaningfulness that gained momentum in the twentieth century. The point is not any more to try to *secure* meaningfulness *at once*, through a criterion of verifiability resting on observability and agreement, which necessarily involve a particular system of coordination, connected to concrete, spatiotemporally located standards. Meaningfulness becomes the *aim* of a necessarily collective, circuitous and time-consuming *enquiry*, where meaningful relations are relations whose statement in a particular system of representation remains valid when it is translated into other systems of representation, according to rules of transformation that characterize equally legitimate systems.

For sure, this can only be a quest, the statement of a task, since the correct structure, and therefore the admissible transformations that characterize it, are not entirely known to us yet. Indeed, every relational structure reconstructed in terms of magnitudes (theoretical concepts) on the basis of empirical relations (measure equations) is bound to depend on the quantities that happened to be accessible to us, and on standards that were easily manipulable. As noted by Nozic, '(t)he development of our understanding of objective facts is a stepwise process, involving mutually modifying knowledge of new facts and of new admissible transformations'.[28] If objectivity is conceived as invariance under (certain) admissible transformations, '(w)e do not begin with an *a priori* criterion of objectivity'[29] – we are not in possession of a complete set of admissible transformations. We know that the magnitudes and equations of classical mechanics reconstructed from measure equations do not all have genuine meaningfulness, even within the classical point of view.[30] As a matter of fact, the choice of length, time and mass as base magnitudes of the SI reflects the constraints imposed by experiment and common practice. We still use these base magnitudes because the determination of physical quantities and relations require the effective exhibition of a coordinate system and the realization of standards accessible to manipulation at the macroscopic scale of human experience, and also because we wish to secure the historical continuity of measurement results.

In this scheme, the referential conception of meaning is recaptured by taking into account the fact that in a scientifically developed society, built on division of work and exchanges, successful reference is conditioned by the requirement that the variety of routes which practical and scientific individuals and communities, driven by different goals, are prone to take when they interact with the world, – thus stirring up various facets of concrete phenomena and coming up with numerous different ways to get hold of secondary magnitudes –, should be coherently coordinated with one another. The referential and the significance conceptions of meaning thus become intimately entangled. We, therefore, have to gradually *work out* what can be considered as commonly warranted, and thus as possibly objective, by looking into combinations of activities which do not involve individual experiences as such, but experiences already intrinsically rooted in social life. The social turns out to be our mode of access to objectivity.

However, that our provisional systems can work and make headway is rendered possible by the gap that exists between the two interpretations – between the magnitude and the measure equations –, a gap that opens up a clearance, a room to manoeuvre and deal with the unavoidable 'imprecisions' of translation – $Transl_1$ as well as $Transl_2$. Interpreted as relations between magnitudes, the equations of physics possess the exactitude of mathematical relations; but as soon as one purports to interpret them as relations between measures, and undertakes to perform actual measurements involving the interaction of an experimental setup with a particular physical system or phenomenon, one is bound to admit that the relations are, in fact, not satisfied. Indeed, the measurement of a concrete quantity always introduces uncertainties. Quantity values can only be estimated, and measures are estimates; these estimates only fully make sense when they are stated with their associated uncertainty. This is, incidentally, a profound reason why neither measures nor quantities can constitute the denotation of the symbols appearing in the mathematical equations of physics. As Giere already argued, the equations of physics cannot refer directly to concrete physical systems and are true only of (mathematical) models.[31] However, far from being a flaw, uncertainty is, on the contrary, a key, constructive feature. Since experiment only gives access to estimates, taking measurement uncertainties into consideration is essential to be able to agree on measurement results and to settle the (provisional) validity of magnitude equations. Uncertainty forms with magnitude an inseparable pair that is indispensable to handle quantities as soon as requirements of accuracy make it necessary to sort out hidden confusions and lay hold of, as well as preserve the room for exchange – between us and the world, between us – in which our construction of common meanings can unfold. In this respect, the magnitude–uncertainty pair appears as the only way to account for the *practical* effectiveness of mathematics in physics.

Acknowledgements

I wish to express my gratitude to O. Darrigol, O. Schlaudt and S. S. Schweber who have commented on different drafts of this paper. I hope that my final version, for which I am wholly responsible, addresses any outstanding queries. My thanks go as well to I. M. Mills for an illuminating discussion. I would also like to thank my language editor, C. Neesham.

5 AN OVERVIEW OF THE CURRENT STATUS OF MEASUREMENT SCIENCE: FROM THE STANDPOINT OF THE *INTERNATIONAL VOCABULARY OF METROLOGY (VIM)*

Luca Mari[1]

Introduction

Measurement is a widespread activity, operated as a critical means to acquire information on empirical phenomena and performed, or wished to be performed, in many (scientific, technical, commercial...) contexts. The information it conveys has a social mark of trustfulness and reliability that is just absent in results of generic numerical estimations, subjective guesses, etc. While the technological and operative side of the measurement endeavour is devoted to enhance these features – variously specified as precision, accuracy (un)certainty... – and is achieved by the appropriate design, set-up and operation of measuring instruments, their justification is a basic task of measurement science: what is there in measurement that, e.g., expert evaluation lacks?

The traditional answer builds upon the Euclidean construction: geometric quantities are paradigmatic of measurable entities, so that the axioms of quantity ground the theory of measurement[2] and foundations of measurement can be spelled out in purely mathematical terms.[3]

Throughout history a parallel stream has emphasized the nature of measurement as an experimental process, and sought in the structure of the process its own justification accordingly.[4] In this perspective measurement is intended to be a common denominator for a multiplicity of subjects, which agree on the social need of relying on the same kind of source of publicly trusted information. On the other hand, exactly the functional role of measurement makes it an infrastructural process, described in different application fields with different, idiolectic lexicons: terms such as 'accuracy', 'precision', 'standard', 'traceability', 'calibration' ... but also 'measurement instrument', 'measurement result' and 'measurement' itself have several context-dependent, and sometimes only implicitly given, meanings, a situation that hinders their shared understanding. Hence, while 'all branches of science and technology need to choose their vocabulary with care [because] each term must have the same meaning for all of its users ... this applies particularly in metrology'.[5]

From this stipulation the *International Vocabulary of Metrology* (*VIM*)was developed, a document on 'basic and general concepts and associated terms' of measurement science. Its very existence, the result of a working group jointly appointed by several international organizations, witnesses the acknowledged importance of establishing a fundamental concept system for measurement. The present paper introduces the *VIM*, and proposes an overview of its history and some driving forces underlying its development (in the following section); then it highlights some significant topics on measurement science that emerge from the latest edition of the *VIM*, published in 2007 (in the section, 'Some Topics on Measurement Science as Interpreted in the *VIM*'), and finally (in the section, 'Some Open Issues for Measurement Science in the Light of the *VIM* Development') mentions some open issues that the *VIM*, but in fact measurement science as such, is facing in its ongoing development.

A Few Notes on the *International Vocabulary of Metrology* (*VIM*)

In the early eighties of the last century four leading international organizations concerned with metrology – the International Bureau of Weights and Measures (BIPM), the International Electrotechnical Commission (IEC), the International Organization for Standardization (ISO) and the International Organization of Legal Metrology (OIML) – agreed 'that there should be a joint action to produce a common terminology [and appointed] to this end a task group ... to co-ordinate the preparation of a vocabulary of general terms used in metrology'.[6] Their shared vision was that 'metrology is a discipline which embraces all the branches of science and technology: the watertight boundaries between the branches are fast disappearing, so it is essential that misunderstandings be prevented when they are caused by different people using the same term to mean different things, or by the use of a completely unfamiliar term'.[7] Their primary challenge was to define a measurement-related concept system in which each term 'at the same time express[es] a well-defined concept and [is] not in conflict with everyday language'.[8]

These two conditions call for a methodological principle able to guide through clashing constraints. Recognizing that unanimity is an ideal state that is not generally achievable, the consensus-based decision making process typical of international standardization was adopted, where the key concept of *consen-*

sus is meant according to ISO/IEC as 'general agreement, characterized by the absence of sustained opposition to substantial issues by any important part of the concerned interests and by a process that involves seeking to take into account the views of all parties concerned and to reconcile any conflicting arguments'.[9] Hence 'consensus, which requires the resolution of substantial objections, is an essential procedural principle and a necessary condition for the preparation of International Standards'.[10] Of course, such a principle would not be pertinent for a terminological dictionary[11] simply reporting the meanings currently attributed to selected terms, and then presenting multiple, and possibly inconsistent or even contradictory, meanings if they are actually found in the relevant scientific literature. But the *VIM*:

> is meant to be a common reference for scientists and engineers – including physicists, chemists, medical scientists – as well as for both teachers and practitioners involved in planning or performing measurements, irrespective of the level of measurement uncertainty and irrespective of the field of application. It is also meant to be a reference for governmental and intergovernmental bodies, trade associations, accreditation bodies, regulators and professional societies. [Hence] this Vocabulary is intended to promote global harmonization of terminology used in metrology.[12]

With these objectives, consistency and absence of contradiction are mandatory conditions, and indeed the definitions in the *VIM* are formulated so that 'the substitution principle applies; that is, it is possible in any definition to replace a term referring to a concept defined elsewhere in the *VIM* by the definition corresponding to that term, without introducing contradiction or circularity'.[13] In this perspective the *VIM* can be intended as a concept system, or even a basic ontology, on the fundamentals of measurement science.

This role evolved in time: from the starting point of the first edition of the *VIM*, published in 1984,[14] two successive editions have been published, in 1993[15] and 2007[16] respectively. What changed in the meantime is the number of organizations involved in the development process – to the initial group four other organizations joined: the International Federation of Clinical Chemistry and Laboratory Medicine (IFCC), the International Laboratory Accreditation Cooperation (ILAC), the International Union of Pure and Applied Chemistry (IUPAC), the International Union of Pure and Applied Physics (IUPAP) – and consequently the scope of the vocabulary, from mechanical and electrical quantities to a broader context in which 'it is taken for granted that there is no fundamental difference in the basic principles of measurement in physics, chemistry, laboratory medicine, biology, or engineering [and the vocabulary also aims at meeting] conceptual needs of measurement in fields such as biochemistry, food science, forensic science, and molecular biology'.[17] Consistently with this standpoint, the third edition of the *VIM* ('*VIM3*' for short henceforth) defines:

[*VIM3*] metrology: science of measurement and its application

hence without principled constraints on the object of measurement.

In this process the fact is remarkable that in 1997 the organizations involved in the development of the *VIM* founded the Joint Committee for Guides in Metrology (JCGM), with the general objective 'to develop and maintain, at the international level, guidance documents addressing the general metrological needs of science and technology, and to consider arrangements for their dissemination', and thus with the specific 'responsibility for maintaining and up-dating the International vocabulary of basic and general terms in metrology (VIM) and the Guide to the expression of uncertainty in measurement (GUM).[18] The very existence of the JCGM witnesses the acknowledged importance of a shared conceptual and lexical ground on the fundamentals of measurement. Moreover, the mentioned changes have been much more than a matter of political moves or lexical fixes: they witness that the fundamentals of measurement science are *a moving target*. The three editions of the *VIM* are an excellent observation point of this evolution.

Some Topics on Measurement Science as Interpreted in the *VIM*

Although only implicitly, the *VIM* in all its three editions assumes the minimal conditions that:

> measurement implies the empirical existence of the measured entity (measurement is not a thought experiment);

> what is measured is not an entity such as, e.g., a table, but an entity related to it, e.g., its length (expressions such as 'measuring a table' are not uncommon, and nevertheless they are just wrong).

Hence, a background ontology for measurement should include at least two kinds of entities:

> entities such as 'phenomena, bodies, or substances', but also individuals, processes, organizations...: let us call them *objects*;

> entities such as length, loudness, extroversion...: let us call them *properties*.

Measurement is about properties of objects (e.g., the length of this table), on which it is supposed to produce information in the form of *quantity values*.[19]

On this basis, while many changes from the *VIM1* to the *VIM3* relate to details, some major components of this evolution can be presented around the following three basic questions:

> what is measurable?

> what is measured?

> what is measurement?

Let us briefly explore them.

What Is Measurable?

As mentioned above, the traditional position assumes measurability as a feature of empirical properties that can be compared with each other in terms of their ratio. The *VIM1* adopted this position by assuming that: (i) quantities are specific properties/attributes; and (ii) only quantities are measurable:

> [*VIM1*] quantity: an attribute of a phenomenon, body or substance, which may be distinguished qualitatively and determined quantitatively

While nominally maintaining these assumptions, and perhaps enhancing the wording of the definition:

> [*VIM3*] quantity: property of a phenomenon, body, or substance, where the property has a magnitude that can be expressed as a number and a reference

The *VIM3* introduces a significant change by extending the set of the properties considered to be quantities so to include those for which only an experimental order can be assessed:

> [*VIM3*] ordinal quantity: quantity, defined by a conventional measurement procedure, for which a total ordering relation can be established, according to magnitude, with other quantities of the same kind, but for which no algebraic operations among those quantities exist

(Note that the very term 'ordinal quantity' would be considered contradictory according to the Euclidean standpoint.) It then defines in a complementary way:

> [*VIM3*] nominal property: property of a phenomenon, body, or substance, where the property has no magnitude

Hence, according to the *VIM3* a property is either a nominal property or a quantity, and a quantity is either an ordinal quantity or a Euclidean quantity for which a unit can be defined. Of course, considering ordinal quantities as measurable entities is a radical break with the Euclidean tradition. The *VIM3* position is indeed that: (i) 'having magnitude' is sufficient for measurability; and that (ii) quantities (including ordinal ones) 'have magnitude' (and conversely nominal properties do not have magnitude and therefore they are not measurable).

This position raises several issues: is the concept 'magnitude' well defined enough, so to support the entire concept system of measurability? Is this extension of measurability to (so-called) ordinal quantities justified? Should the concept be further extended, so as to include nominal properties among the measurable entities? And should the definitions be further revised, so as to take into account the formalization consequent to the so-called theory of scales of measurement that characterizes scale types (nominal, ordinal, etc.) in terms of algebraic invariants,[20] instead of defining them on the basis of the feature 'having magnitude', and maybe deriving property types from evaluation types?[21]

What Is Measured?

The traditional position about the object of measurement is well known: measurement is assumed to convey uninterpreted information on the empirical property in input to the measuring instrument. By accepting this position, so that:

> [*VIM1*] measurement: set of operations having the object of determining the value of a quantity

VIM1 introduced the concept:

> [*VIM1*] measurand: quantity subjected to measurement

In this case, the hypothesis is indeed on the previous, independent existence of the value (that is 'determined') of the considered quantity (the one which is experimentally 'subjected' to the interaction with the measuring instrument). It is to emphasize and to overcome the limits of this assumption that in the last decades several philosophers of science stressed that 'pure or neutral observation-languages' do not exist[22] because 'seeing is a "theory-laden" undertaking: observation of x is shaped by prior knowledge of x',[23] and that finally 'the given [is] acknowledged as taken',[24] so that interpretation is unavoidable, and that this applies also to measurement results.

While clearly refusing a possible radically relativist or constructivist outcome, the *VIM3* acknowledges the lesson, and thus redefines:

> [*VIM3*] measurand: quantity intended to be measured

This is a significant change based on the idea that the measurer is expected to do her best to let the measuring system actually interact with the quantity wanted to measure, but also that the measurement result is attributed to an *intended* quantity, not to the (unknown) quantity with which the instrument actually interacted. The very concept of measurement is redefined accordingly:

> [*VIM3*] measurement: process of experimentally obtaining one or more quantity values that can reasonably be attributed to a quantity

In this case what was assumed to be *determined* is now accepted, in a much less demanding way in terms of ontological presuppositions, as *attributed*, and the quantity to which the values are attributed is the measurand, i.e., an entity that the measurer is expected to define somehow, so as to specify her 'intentions' about it. That measurement results, and in particular measurement uncertainty, depend on definitional issues is a position that was formulated in the *GUM* in terms of an 'intrinsic' uncertainty, 'the minimum uncertainty with which a measurand can be determined, and every measurement that achieves such an uncertainty may be viewed as the best possible measurement of the measurand. To obtain a value of the quantity in question having a smaller uncertainty requires that the measurand be more completely defined'.[25]

On this basis the *VIM3* introduces:

[*VIM3*] definitional uncertainty: component of measurement uncertainty resulting from the finite amount of detail in the definition of a measurand

It also notes that 'Definitional uncertainty is the practical minimum measurement uncertainty achievable in any measurement of a given measurand'.

The idea that the measurand is an entity that must be defined is not new. Consider how already the *VIM1* defined:

[*VIM1*] true value (of a quantity): the value which characterizes a quantity perfectly defined, in the conditions which exist when that quantity is considered

This is streamlined in the *VIM3* as:

[*VIM3*] true quantity value: quantity value consistent with the definition of a quantity

On the other hand the concept of measurand definition is still mainly to be explored (and the *VIM3* does not define it): it paves the way to acknowledge the importance, if not the unavoidability, of *models of measurement.*

What Is Measurement?

The traditional position about the pivotal question of how measurement can be characterized is also well known: measurement is a purely experimental process aimed at discovering the (true) value that the measurand has in relation (and more specifically: in ratio) to a pre-selected unit. This mix (and then alternative support) of ontological realism and epistemological instrumentalism, that founds the naïve assumption that measurement is a 'protocol of truth',[26] is possibly at the basis also of the already-quoted definition of 'measurement' given by the *VIM1*: (i) a determination process whose ideal outcome pre-exists to measurement itself; and such that (ii) the measurand has a single value. The *VIM3* interprets instead measurement (see above) as a process (i) of attribution (ii) of one or more values. The definition is critically grounded on the delicate concept of 'reasonable attribution' of values. How can it be established?

The *VIM3* does not assume a definite standpoint on this issue – thus offering the room for multiple answers, that might be specialized, e.g., in the context of particular fields – but a significant hint may be drawn from the definition of a new concept:[27]

[*VIM3*] measurement model: mathematical relation among all quantities known to be involved in a measurement

This is then specialized as:

[*VIM3*] measurement function: function of quantities, the value of which, when calculated using known quantity values for the input quantities in a measurement model, is a measured quantity value of the output quantity in the measurement model

The meaning of this concept, and thus the actual role of measurement functions, is made clear in some notes accompanying these definitions. First, 'the output quantity in the measurement model, is the measurand, the quantity value of which is to be inferred from information about input quantities in the measurement model'. Second, 'indications ... and influence quantities can be input quantities in a measurement model'. This calls for taking into account the definitions of the involved concepts:

[*VIM3*] indication: quantity provided by a measuring instrument or a measuring system[28]

[*VIM3*] influence quantity: quantity that, in a direct measurement, does not affect the quantity that is actually measured, but affects the relation between the indication and the measurement result

What emerges is a concept of (direct) measurement as a both experimental and mathematical process such that:

- in the *experimental* stage, a measuring instrument is operated, that interacts with the quantity that is actually measured and, usually, some influence quantities, and produces an indication;
- in the *mathematical* stage, a measurement function is computed on the obtained indication value, and possibly correction values and influence quantity values, to produce one or more values that are attributed to the measurand.

Before more thoroughly analysing this idea, let us introduce a simple example to illustrate it. With the aim of measuring a mechanical force (possibly a weight, the measurand), a spring balance (the measuring instrument) is exploited: the application of the force elongates the spring and therefore produces a length (the indication), where such experimental outcome (no quantity values are involved yet) may depend on, e.g., environmental temperature (an influence quantity). It must be now supposed that a value for the spring length l and a value for the environment temperature t can be somehow obtained: these values are then used as arguments of the measurement function m – a properly generalized version of Hooke's law in this case – from which a force value f is computed, $f = m(l, t)$.

The logic at the basis of this structure is straightforward: since a value for the measurand cannot be obtained 'directly' (whatever this means), the measurand itself is transduced to a different quantity – the indication – such that: (i) a value for it can be instead obtained; and (ii) is supposed to be causally dependent on the measurand, so that from the information on the indication some information on the measurand can be inferred. The core component of a measuring system is then:

> [*VIM3*] sensor: element of a measuring system that is directly affected by a phenomenon, body, or substance carrying a quantity to be measured

It is required to behave (i) so to generate an observable[29] output quantity (ii) that is causally dependent on its input quantity subjected to measurement.

It should be noted that this characterization of the experimental stage of the measurement process is purely structural, and therefore independent of the nature of the involved quantities and of the sensor itself: it applies not only to physical quantities and devices but also, e.g., to psychological quantities and tests, where it is supposed that the test outcome – the indication, being it, e.g., the quantity number of right answers to the test items – causally depends on the tested property. In this functional perspective tests are then sensors that transduce competences, attitudes, etc., to numbers of right answers.

If the generic form of a measurement function:

$$\text{measurand} = m(\textit{indication, influence\_quantities})$$

is written by assuming influence quantities as parameters:

$$\text{measurand} = m_{\text{influence\_quantities}}(\textit{indication})$$

it becomes clear that its inverse:

$$\text{indication} = m^{-1}{}_{\text{influence\_quantities}}(\textit{measurand})$$

is the mathematical model of the transduction, where indeed m^{-1} is sometimes called 'observation function' or 'transduction function'. This justifies the idea that the measurement function 'reconstructs' the information on the measurand that has been made available by the sensor through the indication. Hence the mathematical stage of measurement requires the knowledge of the sensor behaviour, as formalized in a transduction function to be inverted in the measurement function.

Even in the case the generic behaviour of the sensor is assumed to be known, each individual sensor is usually characterized by one or more parameters, that must be bound in the measurement function to make it computable. In the example above, the spring is supposed to follow Hooke's law, but its elastic constant has to be determined. This is the critical task for:

> [*VIM3*] calibration: operation that, under specified conditions, in a first step, establishes a relation between the quantity values with measurement uncertainties provided by measurement standards and corresponding indications with associated measurement uncertainties and, in a second step, uses this information to establish a relation for obtaining a measurement result from an indication

The logic of sensor calibration is then:
- one or more measurement standards are given, realizing the quantity of interest and of which a quantity value y_i is known;
- each measurement standard is put in interaction with the sensor, thus producing a corresponding indication of which a quantity value x_i is obtained;
- from the pairs áy_i, x_iñ, together with any possible assumption on the generic behaviour of the sensor (e.g., linearity), the function m^{-1}, and then m, is defined.

This highlights not only that the calibration of an indicating measuring instrument is considered a necessary condition for measurement to be performed by means of it, but also that sensor calibration requires measurement standards that are calibrated in turn, i.e., realize a given quantity with a stated quantity value. As a consequence, measurement results are expected to be provided with the fundamental property of:

> [*VIM3*] metrological traceability: property of a measurement result whereby the result can be related to a reference through a documented unbroken chain of calibrations, each contributing to the measurement uncertainty

In this case:

> [*VIM3*] metrological traceability chain: sequence of measurement standards and calibrations that is used to relate a measurement result to a reference

In other words, the elementary structure of any metrological system is the core component of measurement-related standardization.

Finally, these definitions emphasize the importance of:

> [*VIM3*] measurement uncertainty: non-negative parameter characterizing the dispersion of the quantity values being attributed to a measurand, based on the information used

Indeed:

> When reporting the result of a measurement ..., it is obligatory that some quantitative indication of the quality of the result be given so that those who use it can assess its

reliability. Without such an indication, measurement results cannot be compared, either among themselves or with reference values given in a specification or standard. It is therefore necessary ... characterizing the quality of a result of a measurement, that is... evaluating and expressing its uncertainty.[30]

Measurement uncertainty is generally unavoidable. This is reflected in the change in the definition:

[*VIM1*] result of a measurement: the value of a measurand obtained by measurement

[*VIM3*] measurement result: set of quantity values being attributed to a measurand together with any other available relevant information

In this case the set of values 'together with the relevant information' 'may be expressed in the form of a probability density function'. Measurement uncertainty is then the overall parameter to assess the quality of what a measurement process produces, and as such it is aimed at being compared to:

[*VIM3*] target measurement uncertainty: measurement uncertainty specified as an upper limit and decided on the basis of the intended use of measurement results

In a costs/benefits analysis, if the uncertainty in the measurement result is greater than the previously established target measurement uncertainty, then the obtained information does not have a sufficient quality for reliable decision making based on it; if, vice versa, the measurement result has an uncertainty much less than the target uncertainty, then it can be supposed that an unjustified amount of resources has been devoted to the measurement. Hence, what was deemed to be a purely experimental activity is now understood as a knowledge-based, pragmatic process, in which models play a primary role. This is plausibly the main novelty in the latest edition of the *VIM*: an epistemic turn accepting that even measurement cannot convey 'pure data'. The acknowledged reliability of measurement results – that has been characterized in terms of their objectivity and intersubjectivity[31] – derives from the appropriate design, set-up and operation of measuring systems, including both the experimental and the mathematical stages of the process, not from some 'intrinsic' feature of the measurable entities.

Some Open Issues for Measurement Science in the Light of the *VIM* Development

Measurement science is a moving target, and some of its foundational topics, such as the concepts of quantity, measurand, measurement and measurement result, have significantly changed even in the relatively short time witnessed in the transition from the first (1984) to the third (2007) edition of the *VIM*. Let us mention three of the main open issues for the next steps of this ongoing process.

As defined in the *VIM3*, measurement is largely independent of the algebraic structure of the measurand and the quantity values, and in fact 'measurement implies comparison of quantities or counting of entities', where objects can be both compared with each other and counted also in reference to nominal properties (it is instead in scale construction and instrument calibration that a rich algebraic structure can be exploited, for example by generating the scale through concatenation of units and calibrating the instrument by assuming its linear behaviour). On the other hand, according to the *VIM3* 'measurement does not apply to nominal properties' (and the term 'examination' is introduced to denote the process of value attribution to nominal properties): should the scope of measurement be generalized to nominal properties?

The problem is not a purely lexical one. Invariant transformations depend on scale types (e.g., averaging is not invariant in an ordinal scale), so that most of the *GUM* framework is based on functions that are not invariant for ordinal quantities (from the very basic concept of standard measurement uncertainty, 'measurement uncertainty expressed as a standard deviation'), even though the *VIM3* included ordinal quantities among the measurable entities. The key point for a solution to this problem is, of course, the criterion of definition of what measurement is, whether an experimental or a formal one.

While the concept of mathematical model of measurement is (relatively) well defined, the role of models in measurement is still a subject open to exploration, particularly in reference to the new understanding of measurands as quantities intended to be measured: what is required for the definition of a measurand, and consequently how can definitional uncertainty be evaluated, are delicate issues, in themselves and in view of the definition of 'true value' as 'quantity value consistent with the definition of a quantity'. Moreover, the *VIM3* states that 'there is not a single true quantity value but rather a set of true quantity values consistent with the definition': is this concept of multiple true values correct? And more generally, are the concepts 'measurement uncertainty' and 'measurement error' compatible with each other? (Note that the *VIM3* defines 'measurement error' as 'measured quantity value minus a reference quantity value', i.e., with no reference to true values...)

Is it possible (and reasonable, and useful) to aim at a single, unified and encompassing concept of measurement, that applies both to physical and social/psychological properties and nevertheless is specific enough to maintain measurement distinct from, e.g., generic representation and subjective opinion? And consequently is it possible (and reasonable, and useful) to extend the scope of the *VIM* so to include the treatment of non-physical properties? And, finally, a critical issue is about the infrastructural nature of measurement that has made it a tool for both exploratory and confirmatory analyses in many scientific fields: is it possible (and reasonable, and useful) to build a single concept system of the fundamentals of measurement science that is independent of philosophical positions/presuppositions?

6 CAN WE DISPENSE WITH THE NOTION OF 'TRUE VALUE' IN METROLOGY?

Fabien Grégis

Since the beginnings of error theory and its first developments by Gauss and Laplace, a physical quantity has been characterized by its 'true value', a value that would have been obtained had an ideal perfect measurement been processed. In the second half of the twentieth century, this conception of measurement has been challenged from different perspectives, leading to a position where the concept of 'true value' loses its central place, if not becomes avoidable.

Can we dispense with the notion of 'true value' in measurement? If so, what would measurement be designed for? If not, what precisely would be the philosophical meaning of such a concept, in particular with respect to (scientific) 'truth'? This paper aims at exploring two essential arguments developed against 'true value' in recent metrology texts, in particular the *Guide to the Expression of Uncertainty in Measurement*, henceforth abbreviated to *GUM*,[1] and the *International Vocabulary of Metrology*, abbreviated to *VIM*,[2] two international guidance documents planned to harmonize the terms and practices in measurement science. The first argument consists in pointing out that 'true value' cannot be the proper aim of measurement since it is a forever unknowable ideal. The second argument revolves around the impossibility, in many physical cases, to conceive of a non-unique true value for a given quantity.

It is argued that two different modes of analysis can be distinguished within the critique: an operational mode and a metaphysical mode. By separating these two modes, I defend the idea that metrologists don't actually dismiss the concept of 'true value' as long as they don't adhere to some kind of anti-realism.

The two arguments are presented successively and discussed in relation to their consequences for the notion of 'true value'. The conclusion is dedicated to the relationship between 'true value' and scientific realism.

An Epistemic Turn in Metrology

From 'Error Approach' to 'Uncertainty Approach'

In the traditional approach of measurement that dominated metrology during the first half of the twentieth century, the quality of a measurement procedure was primarily evaluated in terms of its *accuracy*, namely its propensity to produce results with small 'measurement errors'. The latter are defined[3] as the numerical deviation between an actually obtained result and the true value of the measured quantity:

$$\epsilon = x - v \quad (6.1)$$

ϵ = measurement error
x = (known) measurement result (value actually obtained)
v = (unknown) true value of the quantity

In such an approach, measurement is conceived as the determination of an estimate of the quantity's true value with the best closeness of agreement possible. Equation (6.1) shows how measurement error and true value are consubstantially tied. For these reasons, the traditional approach is often designated as an 'error approach' and contrasted with a more recently developed 'uncertainty approach',[4] in which the concepts of 'true value' and 'measurement error' are explicitly challenged. In the 'uncertainty approach', it is argued that their use should be dismissed, as they correspond to ideal concepts, ultimately unknowable.[5] Resting on pragmatic grounds, proponents of the 'uncertainty approach' argue that a proper measurement method, instead of trying to achieve a somewhat metaphysical 'accuracy',[6] should rather only be described through an (epistemic) *uncertainty*, defined as a range of values that are thought be *reasonably* attributed to a quantity.

> The objective of measurement in the Uncertainty Approach is not to determine a true value as closely as possible. Rather, it is assumed that the information from measurement only permits assignment of an interval of reasonable values to the measurand, based on the assumption that no mistakes have been made in performing the measurement.[7]

This aspect is particularly explicit in the evolution of the definition of measurement uncertainty from the first to the third editions of the *VIM*. From 'an estimate characterizing the range of values within which the true value of a measurand lies',[8] it becomes a 'non-negative parameter characterizing the dispersion of the quantity values being attributed to a measurand, based on the information used':[9] the true value has disappeared. To summarize, the difference between the two approaches is primarily a matter of aim. They both share a *fixed* quantity as their common object of inquiry. They differ, however, on whether they aim or not at the true value of this given quantity. This is a change in focus, from the measured object itself to the practical issues related with the application of measurement results. At the same time, 'truth', considered as a somewhat illusory objective, is replaced by the more concrete and accessible *adequacy* with a given goal.[10] This particular evolution participates in a broader movement that we may designate as an 'epistemic turn' in the field of metrology, involving a change of attitude towards the way measurement is considered to be related to knowledge. Another feature of this shift is the transition from frequentist statistics to Bayesian statistics in metrology.

From Frequentist Statistics to Bayesian Statistics in Measurement

The technical machinery of uncertainty analysis has used probabilities since the first developments of a 'theory of errors' in the late eighteenth century. However, if the probabilistic interpretation of measurement results achieved a first maturation in 1827 with what Stigler calls the 'Gauss–Laplace synthesis',[11] metrologists are still faced today with two approaches grounded on incompatible interpretations of probability, namely a frequentist and a Bayesian one.

In the frequentist approach, especially dominant in metrology in the first half of the twentieth century, probabilities are long-run relative frequencies of occurrence of the different possible issues of a repeatable event. The probabilities characterize not a measurement result in itself, but the measurement as a *physical process* of generation of experimental data. The potential outcome of each individual measurement is considered as the result of a random trial from a statistical population. A natural candidate for probability statement is here the statistical distribution of errors generated in a certain given experimental protocol. The idea underlying the error analysis of the measurement process consists in separating what is due to the ideal phenomenon under examination (encompassed by the true value of the quantity) from what is caused by the material and theoretical contingencies of the necessarily imperfect experimental procedure (the 'measurement error'). The statistical analysis of error, given some hypotheses – embodied within a *data model* postulating a general behaviour of measurement errors – enables a statistical inference, formulated in probabilistic terms, about the possible value of the measurand. The frequentist analysis of error shows how the true value acts in the statistical machinery as a regulative ideal that governs how actual values are eventually attributed to the measurand.

However, the frequentist method utterly fails to take into account the so-called 'systematic errors' from a probabilistic standpoint. By nature, this type of error does not vary under repeated measurements. Therefore, it is transparent to any statistical analysis based on frequencies. Proponents of a Bayesian approach in metrology therefore argue that physical probabilities cannot provide a complete probabilistic account of measurement and conclude that epistemic probabilities should be preferred. This position about the role of probability in measurement grew substantially in the latest decades of the twentieth century.

In the Bayesian case, every quantity or parameter of a model – even the fixed values that are systematic errors – may be subject to a probabilistic judgement through a probability distribution, the argument of which being a possible value of the quantity or the parameter, and the corresponding probability density expressing the *degree of belief* granted to the given possible value (in a subjectivist view) or the amount of knowledge that one possesses about the quantity (in an objectivist view. In any case, the subjective character of measurement is generally acknowledged, and even claimed by proponents of the Bayesian approach.[12] Uncertainties and quantity values are inferred from the updating of prior knowledge, given empirical data, by using Bayes's theorem. A measurement result may then take the form of a probability distribution expressing a statement about *the given knowledge (or belief) of a group* about the value of the quantity. The consequence is a displacement of the subject-matter of a measurement result, from the state of the measured object itself to the state of knowledge of the experimenter. For these reasons, the use of Bayesian methods in metrology is at the heart of an ongoing epistemic turn in metrology.

Two Levels of Discussion

To summarize, the evolution of the conception of measurement in metrology in the second half of the twentieth century is mainly characterized by two entangled features:

- A change of focus: the search for metaphysical accuracy, characterized by inaccessible ideals like 'error' and 'true value', is replaced by a formulation centred on the epistemic notion of 'measurement uncertainty'.
- A change in representation: with the growth of the Bayesian viewpoint, a measurement result does not represent the *physical* state of the quantity measured any more. Instead, it formulates a claim about a personal or interpersonal state of knowledge quantified by epistemic probabilities.[13]

Proponents of the epistemic view argue that the traditional objectives are illusory, as it is impossible to formulate a result otherwise than by a statement of present knowledge. In return, defenders of the traditional view, while acknowledging the latter's limits,[14] claim that the Bayesian approach does not enable us to anchor measurement and science in reality.

I will not address this controversy here: rather, my aim is to focus on the change of status that this epistemic turn in metrology generates on the concept of 'true value' in measurement. The latter is illustrated by the following statement by Ehrlich, Dybkaer and Wöger: 'if the true value ... is not knowable in principle, then the question arises whether the concept of true value is necessary, useful or even harmful'!"[15] If the true value is not knowable, what incentives are there to actually use it in scientific theories? I believe that two separate questions can be distinguished concerning the role of the true value in measurement.

(Q1) A metaphysical question: what is the link between 'true value' and *truth* in measurement?

(Q2) An operational question: does measurement need, *as an operation*, a parameter at least similar to 'true value'? In other words, do metrologists actually (need to) *use* anything close to a concept of true value?

I argue that the epistemic standpoint does not in fact dismiss the true value at the operational level, but merely tries to dissimulate it. However, the metaphysical issue remains undecided.

The Operational Problem

Let us answer to the second question first. I will stick here to the definition of 'measurement' given in the *VIM*: a 'process of experimentally obtaining one or more quantity values that can reasonably be attributed to a quantity'.[16] The assignment of a numerical value to a quantity is certainly not *arbitrary*: it is made following certain *rules* deriving from a *data model* embedded into the whole process.

In the frequentist model, the data model describes the probability of each *potential outcome* of the measurement process through a counterfactual long-run frequency of occurrence, would the measurement be repeated infinitely. This probability is *conditioned* on a fixed unknown parameter, the so-called 'true value' of the quantity. The frequentist model then governs the assignment of a value to the quantity by stating that the chosen value to be assigned should be the one that maximizes the probability of occurrence of the empirical data sample actually obtained, *would the true value be equal* to this assigned value, *given the hypotheses of the data model*.

The Bayesian model of data describes the experimenter's state of knowledge through a probability distribution. Any empirical data enables the revision of this probability distribution through Bayes's formula. The likelihood function used in Bayes's formula is itself conditioned on the possible values of the true value of the quantity.[17] As a consequence, it appears impossible, in both approaches, to completely dismiss the *use* of the true value in the value attribution processes, since this concept appears in the *equations* governing these processes.[18] My provisional conclusion is therefore that the concept of 'true value' remains used in current metrology, even if it disappears from the *expression* of the result itself. However, question (Q1) relative to the philosophical meaning of this parameter remains open.

The Metaphysical Problem

Still, even if the notion of 'true value' does intervene in the measurement models actually used by metrologists, one may argue that the denomination is misleading and that the term only designates a useful mathematical tool without any relationship with actual 'truth'. A straightforward way to understand the notion of 'true value' is provided by the correspondence theory of truth attached to a realist account of science: the true value is true in virtue of its correspondence with an actual element of reality. However, following classical philosophical discussions, 'true value' could be interpreted in a weaker sense, as a theoretical term in a theory *adequate* with the results of empirical investigation. In such an empiricist account, all metaphysical reference to 'reality' and 'truth' is rendered unnecessary. In that case, the qualifier 'true value' would indeed be a misnomer, to which could be preferred 'target value',[19] 'theoretical value',[20] or mere 'value', as is the choice made in the *GUM*.[21]

The central claim regarding this issue is the unknowable character of the true value. The emphasis on the unknowable character of the true value displays an empiricist position eager to purge measurement theory from its metaphysical overlay. It is unlikely that this debate may be settled within metrology practice itself: it is rather a question of general philosophy of science. Certainly, the epistemic turn in metrology could be interpreted as an inclination to anti-realism from metrologists. But this does not seem convincing: metrologists and scientists usually tend to stick to a local, moderate realism, such as the one Wimsatt sketches.[22] The status of the concept of 'true value' in metrology still stands as an interesting example of the ramifications of the problem of realism within scientific practice: do scientists apply a specific (personal) philosophy? Do they manage to practice science without any metaphysical or philosophical preconceptions?

Eventually, the empiricist stance described in this section is only one side of the critique of 'true value' that can be found in recent metrology documents. This critique is actually twofold: its other side is directed towards the uniqueness of the true value of a quantity (in a chosen system of units): if a quantity cannot be said to have a unique true value, how could anyone of them qualify as the true value? Although both issues are not clearly separated in recent metrology texts, I believe that they should be considered as two essentially independent problems. The latter relates to a notion recently designated as 'definitional uncertainty', which expresses an idea that has been foreseen for several decades.

Definitional Uncertainty

About Definitional Uncertainty: Definitions

What is 'definitional uncertainty'? It is essential here to stress the importance of the concept of 'measurand' that is the object of enquiry of a measurement. Since the 2008 edition of the *VIM*, it is acknowledged that the measurand is not to be defined as a 'particular quantity (subject to measurement)'[23] but as an *intended* quantity, the 'quantity intended to be measured'.[24] In order to understand the articulation of 'true value', 'measurand' and 'definitional uncertainty', it is useful to quote the following development from the *GUM* in full:

> Suppose that the measurand is the thickness of a given sheet of material at a specified temperature. The specimen is brought to a temperature near the specified temperature and its thickness at a particular place is measured with a micrometer. The thickness of the material at that place and temperature, under the pressure applied by the micrometer, is the realized quantity. The temperature of the material at the time of the measurement and the applied pressure are determined. The uncorrected result of the measurement of the realized quantity is then corrected by taking into account the calibration curve of the micrometer, the departure of the temperature of the specimen from the specified temperature, and the slight compression of the specimen under the applied pressure. The corrected result may be called the best estimate of the 'true' value, 'true' in the sense that it is the value of a quantity that is believed to satisfy fully the definition of the measurand; but had the micrometer been applied to a different part of the sheet of material, the realized quantity would have been different with a different 'true' value. However, that 'true' value would be consistent with the definition of the measurand because the latter did not specify that the thickness was to be determined at a particular place on the sheet. Thus in this case, because of an incomplete definition of the measurand, the 'true' value has an uncertainty that can be evaluated from measurements made at different places on the sheet. At some level, every measurand has such an 'intrinsic' uncertainty that can in principle be estimated in some way.[25]

During the measurement process, the definition of the measurand is *realized* by a *particular* quantity, but *could have been realized* by other particular quantities also consistent with the incomplete definition of the measurand. We may then refer to the definition of true value in the latest edition of the *VIM*: 'quantity value consistent with the definition of a quantity'.[26] Accordingly, the value of each quantity consistent with the definition of the measurand may be said to be a 'true' value of the measurand. As a consequence, a measurand is not characterized by a unique true value, but by a *range* of (equivalently consistent) true values. Thus, what was called 'intrinsic uncertainty' in the *GUM* and in the 1993 edition of the *VIM* and is now designated as 'definitional uncertainty' is defined as follows: 'component of measurement uncertainty resulting from the finite amount of detail in the definition of a measurand'.[27]

Documents like the *GUM* and the *VIM* do not clearly distinguish issues of knowability (discussed in the first section) from issues about definitional uncertainty. However, problems of non-uniqueness of the true value of a quantity were already acknowledged in traditional frequentist approaches.[28] Therefore, it would be preferable to consider knowability and definitional uncertainty as two separate issues.

Estimating Definitional Uncertainty

Measurement uncertainty is quantitatively evaluated through a global 'uncertainty budget' summing up all identified sources of uncertainty. As a component of measurement uncertainty, definitional uncertainty must be accounted for quantitatively. However, the *GUM* does not provide any method of evaluation of this type of uncertainty. In fact, it even immediately gets rid of the problem by stating the hypothesis of 'an essentially unique [true] value'.[29]

There would not be much interest here to enter into a technical discussion about the quantitative evaluation of definitional uncertainty and its integration into the overall statistical theory of uncertainty analysis. However, it is worth pointing out here the *epistemic* nature of definitional uncertainty. Definitional uncertainty *is* an uncertainty because it is related to a given body of knowledge (and not directly to a state of nature): it describes how one *believes*, or knows, the definition of the measurand to be incomplete. If one's body of knowledge evolves, then the definitional uncertainty about a given measurand is likely to change. Definitional uncertainty is not an intrinsic measure of the incompleteness of the definition of the measurand.

In the end, definitional uncertainty is related to the (estimated or measured) width of the range of values consistent with the definition of the measurand, or, more accurately, to the range of true values of the quantities that are believed to realize the definition of the measurand, given an available body of knowledge. This also implies that definitional uncertainty is *not* a consequence of an imperfection of one's body of knowledge (if I ask for a fruit, one may give me an apple or an orange, which equally match what I asked for – this doesn't mean that I don't know how to distinguish between an apple and an orange). On the contrary, a greater body of knowledge will reveal new components of definitional uncertainty to take account of. This marks an important contrast with the ordinary notion of measurement uncertainty which qualifies *limits* of knowledge.[30]

Two Main Issues Generated by Definitional Uncertainty

Definitional uncertainty brings two philosophical difficulties. What is the meaning of a 'non-unique true value'? More explicitly, how can different values for a same quantity be said to be 'true' at the same time? The denomination '*a* true value' once used in the *VIM*[31] to bypass the problem is clearly counter-intuitive.[32] Moreover, non-unique true values conflict with the traditional use of physical equations, which express both a physical relationship between properties and a numerical relationship between the (true) values of the quantities involved.[33] Here, we are confronted again with the two types of issues described in 'Two Levels of Discussion' in the first section: a metaphysical issue about the *truth value* of 'true value' and an operational issue about the role of numerical equations to express physical laws. The next subsection is an attempt to show that the possible answers to these issues are rooted in a separation between the fundamental and the phenomenological.

Reconciling Definitional Uncertainty and True Value: Individuation of Objects and Quantities

The measurement of a given quantity requires us to build a model of the object under measurement,[34] which involves a minimal arbitrary amount of *idealization* and will leave aside some known effects affecting the measurement. In the case of the paradigmatic example of the measurement of the length of a table, the definition of the measurand is conditioned on the idea that the table may be modelled as a geometrical figure (typically a rectangle parallelepiped), the former being an idealization of the table as a physical object.[35] Once conceived of as a geometrical figure, the model of the table involves unique length and width. However, definitional uncertainty expresses how it is impossible to force the reality into the model by measuring a unique value. Thus, the role of definitional uncertainty seems to be at the interface between the physical object and the model.

In many cases, the physical object under measurement is a complex and composite entity that we want to describe as an isolated individual. Let us take the example of the measurement of the length of a pen. A measurand definition would usually be 'the length of the pen' without any more specification. However, a pen occupies more space when held vertically than horizontally, because of gravitational effects (a difference of several micrometres).[36] A more specific definition of the measurand could then be either its 'horizontal length' or its 'vertical length'. Yet, in that case we generally consider that both definitions remain two instances of the *same* quantity in different conditions of observation. What ties these two measurand definitions is the identification of 'the pen' as an individual, a concrete object clearly identified.[37] In general, adding up specifications to the definition of the measurand in order to reduce definitional uncertainty will result in a progressive detachment from the individual itself. Moreover, the pen only exists at a macroscopic scale: once one zooms in into its molecular details, it becomes impossible to distinguish a clear frontier separating the pen from its environment, and thus to even identify an individual. In the end, the identification and isolation of an individual such as 'the pen' is precisely at the cost of a definitional uncertainty: there is a trade-off between individuation and definitional uncertainty. It is because we want to measure the length of 'the pen' in itself, and not to describe a microscopic display of elementary particles, that definitional uncertainty is inevitable at the phenomenological scale. In that case, definitional uncertainty characterizes the fact that, *in our theories or models*, the measurand taken in consideration is a complex individual that presents a *substructure*.

By contrast, the *VIM* states that 'in the special case of a fundamental constant, the quantity is considered to have a single true quantity value'.[38] No definitional uncertainty is attached to fundamental quantities, not because there actually isn't any underlying substructure, but because there exists no *known* substructure: this again highlights the epistemic nature of definitional uncertainty, related to a given body of knowledge and not to the *actual* existence of a substructure. This whole account suggests a double classification of quantities: between phenomenological and fundamental ones on the one hand, and between state variables and constants on the other hand (see Table 6.1). The epistemic nature of definitional uncertainty implies that the status of quantities is not frozen once and for all. Quantities like the mass of the electron (or even Planck's constant) are only fundamental *in the present state of our mainstream theories*: they might not be so in candidate theories such as string theory.

Table 6.1: A suggestion for a classification of quantities

	State variables	Properties ('constants')
Phenomenological	Length of a table	Electrical conductivity of copper
	Mass of a table	Density of water
Fundamental	Quantum state of a given electron	Mass of the electron
		Planck's constant

Definitional uncertainty arises at a phenomenological level, when measurement aims at entities that are not fundamental in our theories. This classification highlights how definitional uncertainty is somehow tied to the issue of reductionism in science, and particularly here through the distinction between fundamental and phenomenological laws of physics. Crucially, phenomenological laws are *approximate* statements – they are only true in a model of phenomena that incorporates idealizations and approximations. Measuring the resistance of a resistor requires a model of this resistor in which it is (for example) assumed to obey Ohm's phenomenological law.[39] What happens here can be put on a par with what happened earlier when the table was modelled as a geometrical figure.

In the models, the phenomenological laws are true and the quantities involved have single values. At the same time, phenomenological laws *are* approximate and phenomenological quantities involved in these laws are associated with a finite definitional uncertainty.

Required Accuracy of Measurement

In the end, the status of the quantities and the equations involved depends on the intentions and the purposes of the experimenter, as underlined by Giordani and Mari:

> Of course, such idealisation is not imposed to the modeler, who, by means of these models, actually decides the concepts (of the object under measurement and the measurand) that she considers appropriate in dependence on her goals.[40]

As Tal explains,[41] measurement uncertainty arises from a progressive 'de-idealisation' of an idealized measurand definition towards the realized one. However, definitional uncertainty differs in this regard. It corresponds to the amount of idealization *not* introduced in the definition of the measurand, but that would have been necessary, given the known physical effects affecting the object under measurement, had one wanted a finer measurand definition, adapted to different purposes. A crucial step resides in acknowledging that, in ordinary measurement, fundamental issues are of *no* interest: the microscopic structure of a pen is of no importance in ordinary measurements about this pen. What matters is the identification of objects at a phenomenological level. This is illustrated by the following quote from the *GUM*:

> In practice, the required specification or definition of the measurand is dictated by the required accuracy of measurement. The measurand should be defined with sufficient completeness with respect to the required accuracy so that for all practical purposes associated with the measurement its value is unique ... EXAMPLE If the length of a nominally one-metre long steel bar is to be determined to micrometre accuracy, its specification should include the temperature and pressure at which the length is defined. Thus the measurand should be specified as, for example, the length of the bar at 25.00∘C and 101,325 Pa (plus any other defining parameters deemed necessary, such as the way the bar is to be supported). However, if the length is to be determined to only millimetre accuracy, its specification would not require a defining temperature or pressure or a value for any other defining parameter.[42]

To conclude, the operational issue generated by definitional uncertainty, that is the compatibility of non-unique true values with physical laws, is defused by noticing that incomplete measurand definitions correspond to phenomenological quantities involved only in phenomenological, henceforth *approximate* laws. Definitional uncertainty then corresponds to a certain 'resolution' at which the phenomenological law is supposed to work, given some intentions represented by what is called in the *GUM* the 'required accuracy of measurement'. The metaphysical issue, though, is more complicated and is rooted in general philosophy of science. It loops back to an issue already brought up in the discussion of the epistemic turn in metrology in the first section. I turn to it in the conclusive section.

Conclusion

The previous study of the critiques addressed against the notion of true value within the field of metrology has unravelled two distinct issues: an operational and a metaphysical issue. I first answer to the operational question: does measurement need a concept of 'true value' in order to be processed? Although I do not a priori dismiss potential alternative attempts to bypass the use of such a concept, I nonetheless argue that none of the main critiques addressed within metrology really establishes that such a use should be avoided and none of them explains how this could be done. On the contrary, the examination of the critique concerning knowability (in the first section) enables us to reconsider how the true value is actually used in value attribution processes, making it a central regulative ideal despite being unknown. Then, definitional uncertainty (analysed in the second section) reveals the importance of the identification of a target in measurement[43] (illustrating how measurement is a goal-driven operation) and shows that the concept of 'true value', at a phenomenological scale, can only be understood by invoking a notion of approximate truth.

If the operational problem is defused, then the critique against the true value is reduced to the metaphysical issue alone. Despite their methodological character, the *GUM* and the *VIM* reveal some inner 'quasi-philosophical considerations'[44] where the position of metrologists about the true value sounds like a variety of empiricism. But is the true value only a tool? The term might be understood in two different ways:

(1) {true} {value of a quantity}
(2) {true value} {of a quantity}

If the true value is understood as 'a value that is true' (case one) then it typically points to the correspondence theory of truth. Otherwise, 'true value' may be understood as being only a denomination (case two) where 'true' is not a qualifier of the value (as was discussed in 'The Metaphysical Problem' in the first section). In this case, the true value might be a useful concept for theoretical construction, but the phrase can only be understood as a whole and has nothing to do with 'true' 'value' taken separately. In the latter case, the true value may then be a statistical parameter, for example the expectancy of some probability distribution, and could for instance be replaced by the term 'theoretical value', in harmony with van Fraassen's empiricist account. This issue is rooted in the opposition between scientific realism and instrumentalism. Concerning measurement and quantities, Michell has defended a realist position.[45] The representational view sees measurement only as an analogy of structures between nature and numbers. Mari tempers these two approaches and sees measurement as being both an assignment (of values belonging to a symbolic world) and a determination (of a state of nature).[46] What appears as particularly important here is to underline that a notion like 'true value' does not necessarily suggest 'the world' to be quantitative *in essence*, but is minimally attached to the truth of scientific claims, scientific theories and physical laws that are expressed through mathematical equations. In physics, quantities are commonly used in equations known to be false: the case of definitional uncertainty shows the importance of a notion of 'approximate truth' if some kind of realism is to be adopted.

Acknowledgements

I wish to thank the editors of the volume, and especially Oliver Schlaudt, for their support. I am grateful to Marc Priel for his kind interest in my work and his useful suggestions. I also thank Luca Mari for his insightful comments during our conversations. I finally thank Nadine de Courtenay and Olivier Darrigol for their careful review on earlier drafts of this paper. The remaining mistakes are mine.

7 CALIBRATION IN SCIENTIFIC PRACTICES WHICH EXPLORE POORLY UNDERSTOOD PHENOMENA OR INVENT NEW INSTRUMENTS

Léna Soler

Introduction

This paper intends to present some of the results of a sociohistorically and ethnologically inspired philosophical research program on calibration in science – a regretfully neglected topic in the philosophy, history and social studies of science. In a previous article co-signed with several members of the research program 'PratiScienS' that I launched in 2007 in Nancy (France),[1] multiple analytical tools have been designed for the purpose of a fine-grained understanding of calibration, and have been applied to *one of the most simple kinds of calibration*: calibration in practices which investigate *relatively well-understood* natural phenomena by means of *already standardized* instrumental devices – for short, UNSI practices (U as understood, N as natural, SI as standardized instruments).[2] In the present paper, the aim is to put the tools previously elaborated at work, for the purpose of an *extension* of the inquiry to calibration in *two other kinds* of scientific practices: practices which explore poorly understood natural phenomena by means of standardized instruments (say PUNSI practices, where PU stands for poorly understood); and practices dedicated to the local invention of new scientific instruments – for example in a given Laboratory (say NewLI practices, where L stands for local and I for instruments). To achieve this aim, a synthetic overview of the relevant analytic tools is first provided (in the following section), and then, these tools are used to offer a (partial) characterization of calibration, on the one hand in PUNSI practices (in the section 'Calibration in PUNSI practices'), and on the other hand in NewLI practices (in the section 'Calibration in NewLI practices'). The ambition is both to get a better grasp of calibration in science and to show the efficiency of the tools designed to this effect.[3]

Analytical Tools for a Fine-Grained Characterization of Calibration

When we scrutinize scientific practices with the aim of understanding calibration, activities candidate to calibration appear to be diversified and at first sight not obviously homogeneous. To cope with such a situation and find a viable way into the 'jungle of calibration', several strategic decisions, associated with the delineation of types and the construction of frames of analysis, have been developed.

Distinguishing Types of Scientific Practices as a Basis to Distinguishing Types of Calibrations

Calibration arguably takes different forms according to the types of scientific practices in which they occur. Accordingly, one of our inaugural constitutive decisions has been to distinguish types of calibration on the basis of a relevant mapping of scientific practices.[4]

The mapping classifies scientific practices by combining essentially two variables and, in a first schematic approach, by making as if each variable worked in a binary way. The first variable corresponds to the *primary target* of the scientific practice under scrutiny. In a binary idealization, this target can be either the natural world, or the means of scientific inquiry (typically, measuring instruments). The second variable corresponds to the *'degree of mastery' associated with the target* of the scientific practice in a given stage of knowledge and technological development. In a binary schematic picture, the mastery is approached as either 'well understood' or 'poorly understood'. Combining the two variables and the two binary values leads to map the field of scientific practices in different types in relation to which activities of calibration show differences that make a difference.

Three of these types will be directly involved below. Two of them, UNSI and PUNSI practices, take the *natural world* as their primary target: their aim is to increase our knowledge about some domain of natural phenomena (a feature indicated by the letter 'N' involved in their acronyms). Both investigate the natural phenomena under interest *with standardized means*, that is, with theoretically and practically well-mastered instrumental devices ('SI'). In other words, actors of these practices are *users* – not designers – of socially well-controlled and entrenched instruments. UNSI and PUNSI practices differ, however, regarding to the degree of mastery of the object they intend to explore. UNSI practices explore phenomena already investigated in the past and relatively well understood as far as their main global features are concerned ('U'). The aim is to go into some details. PUNSI practices, to the contrary, explore poorly understood natural domains ('PU') – typically, subject matters newly constituted as objects of inquiry.

Contrary to UNSI and PUNSI practices, NewLI practices take the *means of scientific inquiry, instruments* ('I'), as the primary target. The proximate aim is the invention of a *new* instrument, even if a more distant aim is the subsequent exploration of nature by means of the new instrument. The specification of the *local* character ('L') is introduced, in order to distinguish, say, the stabilization of a new instrumental prototype in a given laboratory, from the practices of professional metrologists who intend to develop uniform instrumental standards *at a very large scale* – ideally, universal (i.e., internationally accepted) standards. It is necessary to distinguish these two configurations, because in each (at least part of) activities candidate to calibration prove to be notably different. Calibration in professional metrological practices will be left aside in this article, but it is important to keep in mind, all along the path, that they are present so to speak 'in the background', since calibrations in UNSI, PUNSI and NewLI practices recursively depend on established metrological standards.

A Four-Question Frame to Specify Types of Calibration

In order to analyse calibration in *any type* of scientific practice, and to identify the *specificities* of calibration *in each type*, we have introduced a four-question frame, which invites to investigate the four following points.

(i) The *target T* of calibration: What kind of thing can be the *object* of a calibration?

(ii) The *presuppositions Ps* of calibration: what is *taken for granted*, which delimitates what is *not* granted and has to be checked and controlled?

(iii) The *aim* of calibration applied to the target T under presuppositions Ps.

(iv) The *procedure* of calibration: the nature of its structural elements and the kind of stages through which the aim of calibration is achieved.

This four-question frame will be used below to grasp the nature of, and indicate some important differences between, calibrations in UNSI, PUNSI and NewLI practices.

In the present paper, the discussion will focus on the case in which the target is a material *measuring* instrument, and the developments will be illustrated by means of a relatively simple and familiar example, the equal arm balance. The more complex case in which the target corresponds to a *non*-measuring instrument (e.g., an X-Ray source)[5] will not be considered.

Tools for the Analysis of Measuring Instruments Taken as the Target of Calibration

For the conceptualization of the target of calibration identified with a measuring instrument, we have elaborated a general frame, synthesized in Table 7.1 (columns one and two), and applied to the equal arm balance for the sake of substantiation (column three). Below, explanations are provided about the main structural elements of this instrument frame.

Table 7.1: The instrument frame (bold items) applied to the equal arm balance

Generic type	Generic name	Balance
	Mesuranda	Mass of certain kinds of material objects in a given context with a certain degree of precision
	Fundamental scientific principle	Lever principle
	Kinds of quantities typically involved in the mesuranda	Mass
Model	**Name and description**	Equal arm balance
	Scientific scenario	When a massive object O is placed on one pan, it exerts on the end of one arm of the beam a vertical force F_1 whose magnitude is proportional to the mass of O. In order to restore the equilibrium position of the beam, a number of standard masses, exerting a vertical force F_2 equal or to F_1 are placed on the other pan of the balance. The mesurandum (mass of the object O), is then determinable from the instrumental outputs.
	Instrumental outputs	Numbers inscribed on the different standard masses that have been put on the second pan in order to restore the equilibrium.
	Predicted deviations with respect to the optimal working	Drift of the standard masses used ...

For short, a measuring instrument is called a 'measurer'. The function of a measurer is to evaluate quantities of a certain kind. In the example of the balance, the corresponding kind of quantity is mass. The result of the evaluation of a quantity with a measurer usually takes the form of a numerical value associated with a measurement unit and a specified uncertainty.

For the characterization of calibration, it is required to distinguish the type and the token of a measurer. The measurer-type is a *conceptual* object, a *certain kind* of instrumental device. For example: a scale of the type 'balance'. The measurer-token is a *particular material instantiation* of a given type. For example, one of the precision balances constructed by Fortin for Lavoisier.

We call the 'mesurandum' what practitioners *intend to evaluate* in a particular measurement sequence involving a given measurer-token. The specification of the mesurandum *minimally* requires the specification of the *kind of quantities* intended to be measured in the measurements under interest (in our example, mass). But most of the time, it *moreover* involves the specification of *other elements*, such as the targeted degree of precision and/or some environmental conditions (e.g., the external temperature).

When performing measurements with a measurer-token of a given type, strictly speaking, we do not directly obtain the mesurandum. We obtain what we call the 'instrumental outputs' (i.e., measurement results as they would be described by the layman). Once obtained, the instrumental outputs are *convertible*, through a certain scientific scenario (see Table 7.1 for illustration) based on some fundamental scientific principle (for an equal arm balance, the lever principle), in a mesurandum (typically, a numerical value of some quantity). A measurer can thus be viewed as a means to convert certain humanly performed operations (to place an object on one pan of the balance, etc.) into a definite value of a mesurandum.

The involved scientific principle is what defines a *generic type* of measurer. It is what *individuates* a measurer *as one* determined *generic type* different from *another* generic type. Given the fundamental principle that defines a generic type of measurer at the most general level, a multiplicity of different sub-types can be conceived and realized. For example, the lever principle is involved, and differently exploited, in a multitude of sub-types of measurers that can all be used to assess the mass of objects (equal arm balance, Roman balance, Roberval balance, etc.).

Focusing on the 'lowest-level' of such a scale of types, we call the description of a measurer at this lowest-level the *model* of the measurer-type (or, for short, the measurer-*model*). The description of a measurer-model intends to be the characterization of a *real* material measurer-token and its performances, *not* the characterization of an *idealized* measurer.

Regarding the aim to understand calibration, it is useful to distinguish two aspects within the characterization of the measurer-model.

(i) The characterization of the *optimal working*, including an associated uncertainty of measurement. It tells users what is *at best actually obtainable* from a measurer-token generated according to the model (assuming no defaults of fabrication, normal conditions of utilization, etc.).

(ii) The *predictable deviations* with respect to the optimal working. For example, in the case of our balance, a possible drift of the standard masses daily used.

A Simple Exemplar of Calibration in UNSI Practices

To find a tractable route in the jungle of calibration practices, another strategic option has been to start with *one of the most simple configurations*, namely, calibration *in UNSI practices*, and then – since even restricting to UNSI practices, multiple, not obviously equivalent activities can still be identified with calibrations – to construct a 'simple exemplar' of calibration in UNSI practices (cf. Table 7.2). We talk of an exemplar in the Kuhnian sense of a *striking, often encountered, prototypical configuration* (a 'paradigm' in this sense), which does not pretend, *neither to be sufficient* to characterize the diversity of practices candidate to calibration in UNSI practices, *nor to provide a set of necessary and sufficient conditions* for identifying an activity of UNSI practices *with calibration.*

The simple exemplar has two functions. First, it offers a schematic characterization of the kind of activity calibration can be in UNSI practices (in this respect, the simple exemplar is part of our characterization of the *content* of calibration). Second, it works as a compass and an analyser, to the extent that it helps to discuss *other, less simple and more problematic* candidates to calibration in UNSI *and other* practices, these candidates being analysed *in reference* and *by contrast* to the simple exemplar (regarding this role, the simple exemplar is part of our *analytical tools*).

Table 7.2: The simple exemplar of calibration in UNSI practices

Calibration in UNSI practices			
End measurements	Must be specified case by case because they impact on the details of the aim and procedure		
Target	A measurer-token		
Presuppositions	**About the target** (P1) Unproblematic type (P2) Non-defective token		**About the etalon** (if relevant) (P3) Reliability (P4) Adequacy
Aim	A properly tuned measurer-token: to master the obtained/optimal discrepancy for a mesurandum which is as close as possible to the mesurandum of the end measurements		
Procedure	**Tests**	Blank	
		With etalon	
	Operations	Material	
		Symbolic	

For the simple exemplar of calibration in UNSI practices, *the target of calibration* is a measurer-token.

A calibration of a given measurer-token in UNSI practices is never accomplished just for itself, but always with the intention to perform subsequent, more or less pre-determined targeted measurements. We call the latter the *end-measurements* – playing on the two-fold meaning of 'end' (the finality of subsequent reliable measurements and the 'end of an experimental sequence'). Variations of what is intended in the end-measurements can induce variations in the calibration of *one and the same measurer-token* – with important epistemological consequences as we shall see.

Calibration activities in UNSI practices develop under two presuppositions. (P1) The measurer-model is unproblematic: neither the generic type, nor the model of the measurer-token, are questioned. (P2) The measurer-token is not defective: there is no breakdown, no failure. Globally, the measurer works properly – even if it is perhaps not precisely adjusted. The 'distance' between the measurer-token and the measurer-model is not *too* important. Of course (P2) can be questioned in the course of a sequence of actions initially conceived as a calibration. However, if practitioners conclude that the measurer-token is defective, the subsequent operations no more correspond to calibration, but, typically, to a *repair*.

At a general level, the *proximate aim of calibration* for the simple exemplar in UNSI practices can be defined as the achievement of a properly tuned measurer-token. This means to master – i.e., to determine, and if needed to correct – the possible discrepancy between the measurer-token and the measurer-model. More precisely, what has to be mastered is the distance between, on the one hand, the instrumental outputs *actually obtained* with *this* individual measurer-token *at a given time in a given context* (then convertible in the value of a determinate mesurandum, for instance the mass of an object associated with a specified uncertainty), and, on the other hand, the value of the mesurandum that *should have been obtained* in the optimal configuration (that is, *if* this measurer-token as used in this context *actually coincided* with the measurer-model in optimal working). For short, we call the difference between these two items the *obtained/optimal discrepancy*.

This definition, however, must be specified taking into account the *more distant aim* of obtaining reliable results in some *specific* end-measurements. For if, e.g., a very high precision is sought in the end-measurements, the details of the aim of the procedure undertaken to master the obtained/optimal discrepancy will not be the same than if, say, only assessments of orders of magnitude are sought in the end-measurements. Generalizing: the *specific* characteristics of the end-measurements determine *what it means*, for the measurer-token that has to be calibrated, to be '*properly*' tuned. Accordingly, the proximate aim of calibration must be specified as: to master the obtained/optimal discrepancy *for a mesurandum as close as possible to the mesurandum of the end-measurements*.

The *calibration procedures* undertaken to achieve this aim can be decomposed into two logical moments: first, calibration *tests* dedicated to the determination of the obtained/optimal discrepancy; second, according to the conclusions of the tests and if needed, the application of relevant calibration *operations* intended to correct the discrepancy.

Two *species of calibration tests* are commonly involved in UNSI practices: blank, and with etalons. Both tests of the measurer-token under scrutiny must correspond to configurations as similar as possible to the ones of the end-measurements (same environmental conditions, etc.). But blank calibration tests perform measurements *in the absence* of any measured object *of the same kind as the measured object the end-measurements aim to characterize*. Whereas calibration tests with etalons *use measurement standards* as measured-objects. A measurement standard (in French: 'etalon') is an object of already well-known properties. Since the English word 'standard' has a very broad sense which might create ambiguities in the context of our discussion, we prefer to use the French word 'etalon'. Calibration tests with etalons consist in measuring on the etalon, by means of the measurer-token under test, quantities of already-known values.

Calibration tests with etalon in UNSI practices involve two presuppositions about the etalon. (P3) The supposed values of the etalon (certified or assumed on some other ground) are indeed reliable. (P4) The etalon is sufficiently similar to the object under study in the end-measurements: it must be an object of the *same kind*, characterized by the *same kinds of quantities*. In short, we talk of the *adequacy of the etalon*.

Calibration operations can be 'material' or 'symbolic'. *Material* operations correspond to concrete manipulations exerted on the measurer-token, which introduce effective and tangible modifications of the instrument as a particular material body. *Symbolic* operations are diversified, but often correspond to *mathematical corrections* applied to the instrumental outputs actually obtained with the measurer token in the end-measurements.

Now let us see in how far this simple exemplar of calibration in UNSI practices applies to calibration in PUNSI and NewLI practices, taking Table 7.2 as a benchmark, in reference and by contrast to which significant differences can be circumscribed and characterized.

Calibration in PUNSI Practices

Most of the structural characteristic features of calibration in UNSI practices can be transposed to calibration in PUNSI practices, with one important exception: in calibration tests with etalons, presupposition (P4) becomes questionable.

(P4) means that the etalon is 'sufficiently similar' to the measured-object under study in the end-measurements. A 'sufficient similarity' is required, because the etalon *stands for* the measured-objects of the end-measurements: it plays the role of a 'surrogate', and in order to play this role correctly, the etalon, although of course 'numerically different' from the measured-objects, must be 'qualitatively the same'. Otherwise, the result of the calibration test with etalon is *not relevant* regarding the end-measurements. Even assuming that the measurer-token has been correctly calibrated for measurements on measured-objects similar to *this* etalon, the measurer-token in question is still not adequately calibrated with respect to end-measurements involving *notably different* objects and thus notably different mesuranda. The proximate aim of calibration is not achieved, because the requirement, included in its definition, of 'a mesurandum as close as possible to the mesurandum of the end-measurements', is not satisfied.

In UNSI practices, judgements of 'sufficient similarity' between the etalons used in calibration tests and the measured-objects investigated in the end-measurements are, almost always, unproblematic and consensual, *because the objects under investigation in the end-measurements are already well characterized*. Actually (P4)-like presuppositions are usually not formulated at all, and perhaps even not recognized, by actors of UNSI practices. This does not prevent analysts of science from considering that a (P4)-like presupposition tacitly operates in calibration tests with etalons in UNSI practices.

The usually consensual attitude of scientists about the adequacy of the etalon *in UNSI practices*, however, should not conceal the fact that the 'sufficient similarity' condition built in (P4) is not trivial and not regimented by logically compelling criteria. Like any similarity/difference assessment, it involves judgements about what is relevant/irrelevant and important/anecdotal in the scientific configurations under scrutiny. Such judgements are pragmatic and context-dependent appreciations for which no operational universal criteria can be provided, and in some circumstances, they can differ from one practitioner to another. PUNSI practices are one of these circumstances. Since the phenomena under inquiry are poorly understood, issues and dissent can occur about the adequacy of the etalon – leading practitioners to be more aware and explicit about the requirement of a (P4)-like condition. Harry Collins's detailed discussion of the issues raised by the choice of etalons in the episode of the gravitational waves controversy[6] can serve as an illustration of the general points stated above, though mixing the difficulties of PUNSI *and NewLI* practices.

To conclude we can say, leaving binary schematic characterizations, that *the less the object under interest in the end-measurements is well understood, the more the adequacy of the etalon is questionable*. When the adequacy of the etalon is denied, what is questioned is neither the reliability (intrinsic quality) of the etalon (P3), nor the fact that the calibration test with etalon has correctly determine the obtained/optimal discrepancy of the measurer-token *with respect to measurements* involving *this etalon* and other *similar* objects, but, rather, the *appropriateness* of the use of this *type* of etalon *for* the acquisition of knowledge *about the still poorly understood phenomena under inquiry in the end-measurements*, or put differently, the fact that the mesurandum of the calibration test is close enough to the mesurandum of the end-measurements. If the 'near identity' between the etalon and the poorly understood measured objects is challenged, the calibration test does not say anything about the ability of the measurer to detect the hypothetical new phenomena and, providing they exist, to offer meaningful information about them.

The rejection of (P4) can turn to be specified in two ways.

(i) *Another* type of etalon can be found, which is taken to be adequate – able to provide a *physically informative* calibration test *regarding the mesurandum of the end-measurements*. Collins's account of the gravitational waves episode offers an illustration. Some of Weber's critics first performed an 'electrostatic calibration' of their instruments, but because the corresponding calibration test was questionable *as an adequate surrogate* to experiments *with gravitational waves*, one physicist designed an alternative 'spinning bar calibration', claimed to be 'more appropriate with something like a Weber resonant antenna'.[7]

(ii) It is concluded that the *model* of the measurer-token is not appropriated for the investigation of the phenomena targeted in the end-measurements, and that *other* models of instruments should be used or designed. Weber's position in the gravitational waves controversy provides an illustration. If his competitors fail to detect the gravitational waves Weber claims to have detected, it is, according to Weber, because their experimental means are not appropriate to detect the *kind of phenomena* at stake, whereas his own device is. Here, the source of the problem is the adequacy of the choice, not just of the etalon, but of the model of instrument. If the measurer-model is unable to detect or reliably characterize the kind of phenomena under inquiry, any etalon is inadequate, and any calibration test is irrelevant as a means to warrant the reliability of the measurer-token *regarding the mesurandum of the end-measurements*. The proximate aim of calibration cannot be achieved, because the model of the measurer does not offer the possibility to obtain a mesurandum *close enough to the mesurandum of the end-measurements*.

Taking possibility (ii) into account reveals that a fifth presupposition, not explicit in Table 7.2 but added in Table 7.3, is involved in calibration in UNSI practices: (P5) Adequacy of the measurer-model. (P5) means that the model of the measurer-token under test is able to give reliable information about the kinds of quantity, and more generally the mesuranda, involved in the end-measurement. (P5) is totally unproblematic in UNSI practices – and so trivial that it is easily unnoticed.

Taking the simple exemplar in UNSI practices as a compass and an analyser, we have been able to specify *exactly what varies* (see Table 7.3) in the activity of calibration, *and why*, when turning to PUNSI practices – what varies when already standardized or sufficiently well-mastered instruments are applied to *poorly* rather than *well*-understood phenomena, *all other things being the same*.

Table 7.3: Calibration in PUNSI practices

End-measurements	Must be specified case by case because they impact on the details of the aim and procedure, *and possibly on its adequacy*	
Presuppositions	**About the target** (P1) Unproblematic type (P2) Non-defective token <u>(P5) Adequacy of the</u> <u>measurer-model</u>	**About the etalon** (if relevant) (P3) Reliability ~~(P4) Adequacy~~

Italicized items: differences with calibration in UNSI practices
<u>Underlined item</u>: a common presupposition of the PUNSI and UNSI practices, concealed in the UNSI case but revealed by the analysis of the PUNSI case

Calibration in NewLI practices

By the local invention of an instrumental device, I mean the intellectual, material and practical aspects of the process through which a new instrumental prototype is designed, investigated, and finally locally stabilized (for example by a few number of actors pertaining to a given laboratory). More or less innovative instrumental prototypes can be considered, from the case of a slightly revised version of an already well-mastered model based on unproblematic scientific principles, to the case of a revolutionary generic type based on some not fully understood scientific principles. Whatever its degree of innovation, the corresponding process, including its end point, is often described in terms of calibration.[8]

What, exactly, is (or can be) calibration in this case?

General Remarks on the Aim, Target, Presuppositions and Procedure of Calibration in NewLI Practices

In practices dedicated to the invention of new instrumental devices, as in any other type of scientific practices, calibration *minimally* refers, at the *most general level*, to the achievement of a reliable instrument-token (typically a measurer) regarding some mesuranda. The latter achievement can thus be equated with a general definition of the aim of calibration. When such an aim is achieved, the corresponding measurer can be said to be calibrated. At this general level, the specificity of the NewLI case is, compared with the UNSI case, that the potential routes to a successful calibration are much more open, and that the *very possibility* of a *successful* calibration cannot be warranted (whereas in UNSI practices, there is no doubt that measurers *of the models involved* can indeed be successfully calibrated). Accordingly, at the most general level, we can define the aim of calibration in NewLI practices as follows: to decide *whether or not* a given instrumental prototype, as conceived and embodied at a given stage of the inquiry, has the ability to provide reliable information about some mesuranda.

To go further, we must have a closer look at the process through which such aim may be achieved. At each stage of the process of invention of a new instrument, practitioners struggle to provide answers to the following questions: how to materially transform and/or reconceive the measurer-token under development, to be in a position to convert its instrumental outputs in trustful values of the kinds of quantities involved in the measuranda under interest? What are, in the instrumental outputs obtained with the prototype, manifestations of the natural reality on the one hand, and artefacts produced by the measurer-token as used in this context on the other hand? In the process, both the measurer-token as a *material* object, and the measurer-model as a *conceptual* reality, can, and usually are, modified. Most of the time, the identity and stabilization of the measurer-model and of the measurer-token evolve simultaneously, and are mutually adjusted, through an open-ended iterative process of co-maturation. Successive versions are attempted, until a satisfying stabilization point is possibly achieved.

Thus, at the end of a successful invention process, what is finally stabilized and taken as reliable is, simultaneously, a material instrument-token and something akin to a conceptual measurer-model. *Something akin to* a measurer-model, because the local invention of a new instrument does not always end in a well-articulated, complete and detailed theoretical explanation of the instrument. As diverse scholars stressed, to be able to use an instrument-token, it is not necessary to master *all the theoretical details* of the scenario through which the instrument converts human operations in the values of some quantities.[9] However, a minimal representation of the way the instrument works – and hence a 'theory' in this non-technical sense – even possibly rough and fragmentary, is required to be in a position to believe that the instrument-token is reliable for some purposes. I will continue to use the term 'model' to name the conceptual representation associated to the instrument, including for cases in which this representation is very sketchy.

Relying on the previous analyses, we can specify the target of calibration in NewLI practices: The target is simultaneously the measurer-token *and the measurer-model* (and not just the measurer-token as in UNSI practices).

Presuppositions about the target also differ from the UNSI case. (P1) does not hold: the model (and in innovative cases, possibly also the generic type) *is problematic*, since it is part of what has to be designed, investigated and stabilized. As for P2 (not-defective token), it is irrelevant: the issue of the proper working of a measurer-token (i.e., whether or not the instrument is defective), is the issue of deciding whether or not the measurer-token is globally conform to the prescriptions of its model, and this only makes sense under the condition that a taken-to-be-reliable model has already been stabilized. I will not attempt to specify what the presuppositions *positively* are in this case.

The aim of calibration in NewLI practices can now be *specified* as follows: (a possibly unique token of) a material object conceived as a particular instantiation of a (more or less detailed and complete) taken-as-reliable measurer-model. Part of the aim is to define – or at least sketch – a model of the instrument, including the definition of a proper working of the prototype. Once the aim has been achieved, the resulting instrument can be *used* for some purposes, and then, calibration *as defined in the UNSI case* enters in play – i.e., the search for an instrument which is properly tuned with respect to some end-measurements, assuming that the instrument works properly. Still more concretely, the aim of the calibration *tests* performed in the process of invention of a new instrument is – to formulate the point in a way intended to facilitate the comparison with UNSI practices – to evaluate, and if needed to reduce as much as possible, the discrepancy between, on the one hand, an interpretation of the instrumental outputs actually obtained with the measurer-token, and, on the other hand, some *expected* measurement results that should have been obtained if the measurer-token was reliable. For short, say to master the obtained/expected discrepancy.

Let us now consider such calibration tests more closely. Two widespread and typical cases will be (partially) discussed below.

Calibration Tests with Etalons

The first case of calibration test will only be briefly discussed, because calibration tests with etalons have already been characterized for the UNSI case, and because the principle of the test is the same in the two cases, namely, to perform measurements on objects of already-known properties with the measurer-token under test, so as to evaluate the discrepancy between the obtained and the expected results. I will concentrate on *differences* between the *aims* of calibration tests with etalon *in NewLI and UNSI practices*.

Considered at a fine-grained level, the aims differ in the two cases. In NewLI practices, the aim is to decide whether or not a scientific scenario, based on some model of the measurer-token, can be found, that enables to convert the instrumental outputs obtained, with this measurer-token as the result of some human operations involving the etalon, into the known properties of the etalon. By contrast, in UNSI practices, such a scientific scenario is available and *taken for granted* from the start (this is built in (P1)). Taken for granted that *in optimal working*, the instrumental outputs of a measurer-token of this model *can be reliably converted into the known values of the etalon*, and assuming moreover that the measurer-token is not 'too far' from its model (P2), the aim is to determine the magnitude of the possible slight deviation of the measurer-token from the optimal working.

When, in NewLI practices, the instrumental outputs of the measurer under test prove to be interpretable in terms of results that are *close enough* to the expected results – here, close enough to the known properties of the etalon – the calibration test is successful, and the instrumental prototype is calibrated in the sense that this measurer-token, viewed as an instantiation of some measurer-model, is assumed to be able to provide sound information about the kinds of quantities measured on the etalon. Note that strictly speaking, calibration claims of this kind should not mean more than 'the measurer has been calibrated *regarding the etalon involved in the test*'. In practice, however, when a calibration test with etalon performed on a new instrument is successful, practitioners are confident that the instrument under test is reliable *not only for the etalon used*, but for a larger class of objects *similar to the etalon*.

When the results obtained with the measurer-token are *completely out of range* regarding the results expected with the etalon, the calibration test is unsuccessful, and the measurer is not (or has not yet been) calibrated (further work might be needed to examine if calibration can be achieved). The nature of the discrepancy between the expected and obtained results can sometimes suggest possible improvements of the material and/or conceptual prototype. The corresponding actions can be viewed as (attempts of) calibration operations.

Calibration Tests of a New Instrument against a Different Instrument Taken as a Reference

This calibration test uses, as a reference point, not an etalon identified with a measured-object of well-known properties, but a second already well-mastered measurer, say the 'reference-measurer' (we could also talk of a standard-measurer or a calibrating-measurer). The principle of the calibration test is to make measurements on the same measured-objects, on the one hand with the reference-measurer, and on the other hand with the measurer under development, and then, to compare the interpretations of the instrumental outputs obtained with the two different measurer-token, *taking the reference-measurer as a cornerstone*. An illustration is provided by the oft-cited historical episodes in which the light microscope played the role of reference measurer to test new models of electron microscopes.[10] The procedure of using a new, poorly mastered instrument against a different, already well-mastered instrument taken as a cornerstone seems to correspond to the most common understanding of the term 'calibration'.

In order to play its role of reference, the reference-measurer involved in the calibration test must satisfy some conditions. Below is a not exhaustive list of presuppositions about the reference-measurer.

(i) The reference-measurer is an already well-mastered and largely unproblematic measurer. In other words, (P1) holds: the measurer-*model* which works as a reference is (sufficiently) unproblematic.

(ii) The reference-measurer is 'sufficiently' different from the new measurer under test. This delicate requirement is often expressed in terms of 'independence': the two measuring systems must be 'sufficiently' independent from one another. Such judgements are first of all judgements about the two *models* of the measurers at stake. They mean that the two models are not based on the same scientific principles or do not use the same theoretical ingredients.

(iii) (P2) holds: the measurer-token which works as a reference must not be defective.

(iv) In most cases, the reference-measurer-token is calibrated in the sense defined for the simple exemplar of calibration in UNSI practices (properly tuned). In other words, the reference-measurer-token coincides with the optimal working as defined by the reference-measurer-model (or if not, practitioners know the magnitude of the discrepancy and are able to correct the obtained instrumental outputs accordingly). The 'properly tuned presupposition' is the analogue of the reliability of the etalon (P3) for calibration tests with etalons.

(v) The reference-instrument provides information about mesuranda that are *sufficiently similar to the ones intended to be determined with the new instrument* – i.e., minimally, information about the *same kinds of quantities* (it would have no sense to test, say, a model of thermometer against a model of balance). Let me talk of the adequacy of the reference-measurer (with respect to the mesuranda intended to be determined with the new measurer). This presupposition is the analogue of (P4), the adequacy of the etalon for calibration tests with etalons in UNSI practices – though in the latter case, the adequacy is evaluated with respect to the mesuranda *of the end-measurements*.

Under these presuppositions, the aim of calibration tests is to check whether or not the instrumental outputs of the new measurer can be interpreted as *delivering the same message*, or at least as *not being in disagreement with* the results (with the standardly interpreted instrumental outputs) obtained by means of the reference-measurer. To illustrate with the example of microscopy, the task implies to decipher, in the instrumental outputs of the electron microscope, what manifests properties of, say, bacteria, and what corresponds to instrumental artefacts. The issue is whether or not it is possible to establish a mapping between, on the one hand determined characteristics of the objects under study, as revealed by the interpreted pictures delivered by light microscopes and taken as a fixed reference, and, on the other hand, aspects of the pictures delivered by electron microscopes.

If such mapping can be found, we have a successful calibration test. The instrument under test (e.g., the electron microscope) has been calibrated against the reference-instrument (e.g., the light microscope). It means that the electron microscope is able to give reliable information about some mesurandum (e.g., some properties of the bacteria involved in the calibration test), and that something like a proper working of the new electron microscope has been stabilized. Once again, strictly speaking, we should only say that the electron microscope has been calibrated against the light microscope *with respect to* the bacteria under study, or more precisely with respect to the kinds of quantities involved in the calibration test that have been concluded to be in agreement with the pictures delivered by electron microscopes. But in practice, calibration claims often convey the idea, or at least the hope, that electron microscopy is a relevant and reliable means to gain knowledge about the microstructure of *other similar* bacteria and objects.

Calibration Operations in NewLI Practices

In NewLI practices, as one can see relying on the previous analyses, the vocabulary of 'calibration' is employed in a broader and more encompassing sense than in UNSI practices. Taking into account this more 'permissive' use, calibration *operations* in NewLI practices can encompass all intellectual and material accommodations which make the instrumental outputs of the new instrument *cohere* with what is already taken for granted in the present stage of scientific development – where 'what is already taken for granted' might be provided either by an already well-characterized measured object working as an etalon, or by an already well-mastered instrument working as a reference.

Conclusion

As a conclusion, let me note that the case of calibration of *poorly mastered instruments* for the sake of the investigation of *poorly understood phenomena* can be easily grasped using the previous characterizations of PUNSI and NewLI practices, as a mix of the specific difficulties involved in each.

8 TIME STANDARDS FROM ACOUSTIC TO RADIO: THE FIRST ELECTRONIC CLOCK

Shaul Katzir[1]

In the aftermath of the First World War researchers at the British National Physical Laboratory (NPL) and the American Telephone and Telegraph Company (AT&T) contrived and constructed a new kind of frequency standard. Based on electronically maintained vibrations of tuning forks, their standards suggested hitherto unknown accuracy in measuring high frequencies. Moreover, to ensure the accuracy of these measuring standard they designed another novelty – the first electronic clock. Although the mechanical vibrations of the tuning fork controller continued to determine its pace, designed for and based on electronic technology, the new clock symbolized and helped the turn from mechanical to electronic technology. The transformation to the electronic world is manifested even more clearly with the electronic successor of the tuning fork clock – the quartz clock, in which the hidden electro-mechanical vibrations of the crystal resonator replaced the tuning fork controller.

Notwithstanding the novelty of the new electronic tuning fork standard, it was based on an established tradition of exactitude in the mechanical research of acoustics. The idea of a tuning fork clock, albeit not an electronic one, also originated in that tradition. This article examines how and why the measuring methods developed within the study of nineteenth-century acoustics were taken for twentieth-century electronic technology. In following the history of the electronic tuning fork it points at three major factors that made this transformation possible. First was the rapid development of electronic radio techniques in the 1910s. Yet, no one designed electronic tuning fork techniques just because new means became available. Researchers, rather, devised new methods to answer particular technological needs, as they were defined by military, commercial and governmental organizations in relations to their goals and the current state of technology. Here, I show that the needs and, even more so, the goals were diverse, and that their particularities shaped the technological solutions. Accuracy became imperative for World War I research on improved wireless devices and ballistics in France and Britain. Yet frequency measurement standards were developed for governmental and commercial civilian needs at the war's aftermath. To coordinate telecommunication within their jurisdiction, Governments needed means for measuring frequencies of radio transmissions, which allowed allocating each to a given wavelength band. AT&T needed a common scale of frequencies to prevent clashes in its elaborated telecommunication system and to allow transferring signals between different methods in its use. As in other cases, accurate measuring standards served for coordinating activities within large networks, where agreement about magnitudes was essential for the functioning of the network as a whole.[2] These contemporary needs were the second factor. A third factor was the strong scientific background of the researchers involved. Mobilized to the war effort and recruited to the novel governmental and industrial research laboratories, physics graduates carried out the research on the new frequency measurement methods. Their academic and practical experience in physics enabled these physicists and engineers to adopt not only methods and ideas but also the ethos of exactitude from nineteenth-century acoustic to the electronic technology of the early twentieth century.[3]

Abraham and Bloch

At the eve of the First World War, the tuning fork was the central device in exact measurement of frequencies for music and scientific ends alike. Since 1834, the tuning fork suggested a portable and thus useful 'measuring stick' for frequencies, calibrated by a few audial and visual methods.[4] Moreover, tuning forks were incorporated in a few mechanical and electromagnetic instruments that allowed maintaining their oscillations and their use for measuring electromagnetic oscillations. These instruments, however, could not be directly used for wireless since tuning forks could not vibrate at the high frequencies (of thousands of cycles per second – Hertz) commonly in use in radio communication. A way to apply the tuning fork for measurements in the new realm was suggested by a new device – the multivibrator.

The 'multivibrator' originated in the French military research on radio communication. Henri Abraham, a physics professor at the *École normale supérieure*, and Eugène Bloch, an active physicists and a teacher at a *lycée*, invented the device. Recruited to help research at the military radio-telegraphy, Abraham and Bloch examined some anomalies with the behaviour of valve amplifiers in use. These amplifiers were based on the novel 'triode': an electronic vacuum valve with an additional third electrode, called 'grid', which controls the electric discharge from the 'cathode' to the 'anode'. Applied for emitting and receiving electromagnetic waves and for amplifying currents, triodes (also known as 'audions') became the basis for wireless communication during the war. They have kept their central role in electronics until the advent of the transistors in the early 1950s. Before the war, AT&T developed triodes, originally invented for transmitting voice over wireless, as amplifiers for telephony, the main business of the corporation. Other researchers refined and applied the improved tube for radio, where it opened up new possibilities. French researchers, however, had not entered the development of triodes until the beginning of the war. Yet, during 1914–15, enjoying American and German knowledge, actual prototypes and the expertise of Abraham and other civil scientists and engineers, the French military radio-telegraphy devised and produced state-of-the-art valves and multi-valve amplifiers. Its researchers continued to spend much effort in their improvement.[5]

Examining such multi-valve amplifiers during 1916–17, Abraham and Bloch 'noticed irregular discharges in these devices'. On further examination they identified the cause of the discharge in the way the triodes were connected to each other. At this point they realized that they could employ this interfering discharge, which they originally tried to remove, to determine high frequencies used in radio, a basic step for further measurements in radio research. Oscillating tens of thousands times in a second, much higher than vibrations hitherto used and studied, radio waves posed a new challenge for frequency measurement. The war research sharpened the challenge as it brought with it significant advancements in radio, and scientists with an interest in exact measurements to its study. Contemporary wave-meters, however, were insufficient for the new requirements as they hardly exceeded an accuracy of one per cent.[6] Abraham and Bloch, therefore, seized the new effect they found in unconventionally connected triodes to improve the accuracy of frequency measurements.

Consequently, the two physicists changed the goal of their research. Instead of using the triodes for amplifying alternate current without changing its frequency as they initially tried, they used triodes to multiply known frequencies. Rather than eliminating the discharge as needed for their original goal, they augmented it by coupling the triodes in a new manner: connecting the grid of each triode to the anode of the other triode through a capacitor (Figure 8.1). This connection generated periodic discharges between the valves. The new device could generate electric oscillations in many frequencies that are exact integer multiplications of the input frequency. Following the way they appeared (and heard) in acoustics, students of periodic phenomena dubbed such oscillations 'harmonics'. According to Abraham and Bloch their new device 'is true extraordinary rich in harmonics reaching an order of 200 or 300. [Therefore] we gave this device the name *multivibrator*, which reminds this remarkable property'. Due to its production of high harmonics, the multivibrator allowed generating high frequency oscillations from low frequency oscillations. Since precise and stable low frequency vibrations were easier to produce, the method offered high frequency oscillations of higher stability than could be produced by direct means. These oscillations could be the basis for a new frequency meter for radio frequencies.[7]

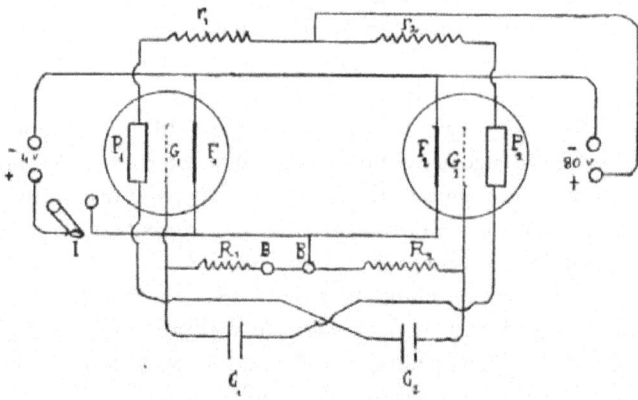

Figure 8.1: The multivibrator. The two circles are the triodes, where F is the cathode (filament), P the anode (plate) and G the grid. Notice that the grid of the left triode (G_1) is connect to the anode of the right triode (P_2) through a capacitor (C_1) and vice versa. The figure includes also resistors and direct current power sources.

Source: Abraham and Bloch, 'Measure en valeur absolue', p. 214.

Abrahams and Bloch chose the tuning fork as the reference for the low frequency multiplied by the multivibrator. Thereby, they employed its mechanical precision for the novel electronics. During the nineteenth century scientists, like Ernst Chladni, Jules Lissajous, Herman von Helmholtz and Lord Rayleigh and instrument makers like Rudolph Koenig, studied and improved the stability, purity and exactness of the tone produced by the U-shape tuning fork in common use by musicians. In their hands it became the most precise device to measure frequencies. They also employed it to regulate the production of sound pitch and as a timer for short intervals.[8] By employing the tuning fork, Abraham and Bloch connected the new field of electronic oscillations to the established tradition of physical research on mechanical vibrators.

Yet, the multivibrator could not be directly connected to the mechanical vibration of the tuning fork, but only to electric oscillations. Thus, Abraham and Bloch connected the multivibrator to a circuit whose frequency was equal to that of the tuning fork. To ensure the equality, they transduced the electro-magnetic oscillations of the electric circuit into mechanical wave, i.e. sound, through a telephone receiver, a common tool in early electronic laboratories. Comparing the sound heard through the telephone with the tuning fork's pitch, they reached a deviation of no more than one-thousandth between the two, as higher differences produced audible beats. This mechanical-electrical transduction reduced the accuracy in one order of magnitude, since the frequency of their 1,000Hz (near Soprano do) tuning fork was stable to within one in ten thousand. As common in the acoustic tradition, the latter frequency was determined through a comparison to a clock, since frequency is the inverse of time, and time was determined more exactly than any other magnitude. Using a known method, Abraham and Bloch, recorded the vibrations of the tuning fork on a photographic film and compared them to the second beats of an astronomical clock. In this method, they connected the tuning fork to a small synchronous motor to reduce the rate of its marks on the film. The harmonics produced by the multivibrator-based tuning fork could be served as a high frequency wave meters, of about one order of magnitude more precise than those in common use.[9]

Shortly thereafter, researchers dispensed with the audial comparison, which reduced the precision, and incorporated the tuning fork into an electronic circuit. Interestingly, at about the time they developed the multivibrator, Abraham and Bloch invented a triode circuit that oscillated at the frequency of an embedded tuning fork. Still they did not suggest connecting this circuit to the multivibrator. In their method, the mechanical vibration of a tuning fork (or a pendulum) induced alternating magnetic force and consequently an electric current at its period in induction coils, placed in proximity (Figure 8.2). Simultaneously, alternate electric current in the coils induces a changing magnetic field, which exerts a force on the magnetic tuning fork. At suitable frequencies this alternating force maintains the mechanical vibration of the tuning fork against damping. Since the triode electric circuits were flexible enough to oscillate in a range of frequencies, the mechanical vibration forced the frequency of the electric current in the circuit to its own period.[10]

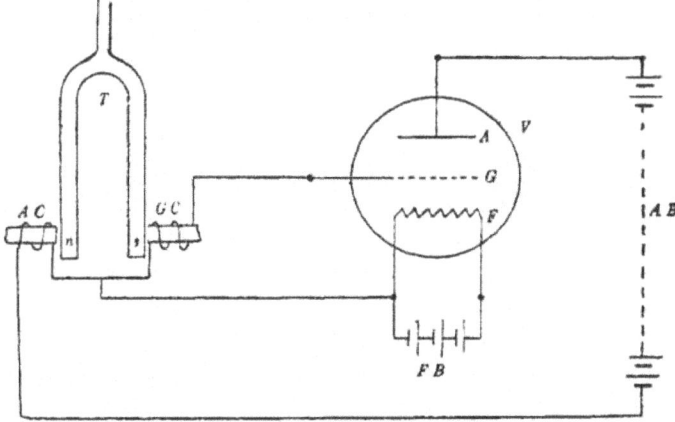

Figure 8.2: A triode-maintained tuning fork. T is a permanently magnetized tuning fork, AC and GC are two induction coils, connected to the triode (V), one to its cathode (F) and the other to its grid (G), A is the Anode.

Source: S. Butterworth, 'The Maintenance of a Vibrating System by Means of a Triode Valve', *Proceedings of the Physical Society of London*, 32 (1920), pp. 345–60, on p. 345.

Originally, Abraham and Bloch developed the pendulum-controlled triode circuit to maintain steady alternating currents of a few cycles per second, in order to amplify low frequency oscillations for unspecified needs of the military radio-telegraphy.[11] Coupling a pendulum to an electromagnet was common in electric pendulum clocks.[12] Abraham and Bloch's novelty lay in coupling the mechanical vibrator to an electronic circuit based on the new triode valves. Once they had a low frequency circuit based on the pendulum, extending their method to the tuning fork was a simple step. Beginning with Lissajous and Helmholtz in the 1850s, scientists and instrument makers had already suggested magnetic coupling of vibrating tuning forks to electric circuits. The connection of the tuning fork to electromagnetic systems was, thus, common within a tradition of exact measurement. Moreover, in 1915, Amédée Guillet, a lecturer at the Sorbonne, devised the first electric tuning fork chronometer, based like Abraham and Bloch's device on magnetic coupling. Yet, unlike Abraham and Bloch, Guillet did not use electronic valves, but employed, instead, a carbon microphone as a current rectifier. His device was exact only for short intervals, but probably satisfied its role for laboratory measurements. As a student of Gabriel Lippmann, Guillet had a strong interest in metrology especially of electromagnetic units but also of time. His chronometer incorporated a method for keeping the vibrations of tuning fork, which he had previously developed for exact measurement, with a new method for counting the electric oscillations. Since Guillet published his suggestion, and moved in the same circles as Abraham and Bloch, they must have heard about his method.[13] Yet, they did not need Guillet's particular suggestion to know that one can couple tuning forks to electric circuits.

Eccles, Jordan and Smith

The war research on improving wireless led also the Brits William Eccles and Frank Jordan to invent a similar method of 'sustaining the vibration of a tuning fork by a triode valve', independently. Interestingly, Eccles and Jordan developed the method to measure the magnifying power of valve amplifiers, a goal for which Abraham and Bloch designed the multivibrator. Later, Eccles traced the origins of this work to commercial motivations. Namely, in 1914, he suggested the use of tuning fork to circumvent a patent of the German firm Telefunken, which allowed using only one amplifier for both sound and radio waves regardless of the differences in their frequencies. It reduced, thereby, electric consumption and expenses on additional amplifiers and thus threatened to dominate radio. Eccles tried unsuccessfully to utilize the harmonics of a tuning fork to reach both audial and radio frequencies from one vibrator. He had already been a notable expert on radio communication both in its theoretical understanding and in its practical uses. A reader at University College London, he had carried out many studies on a range of radio topics from wave detectors to the ionosphere.[14]

During the war, Eccles consulted a few military arms on radio. For the needs of his new clients, he modified the tuning fork circuit, most importantly by incorporating a triode. With the help of Jordan, an 'electrician' and a 'lecturer of physics' at the City and Guilds College in which Eccles became a professor of applied physics and electrical engineering in 1916, he 'found that a generator of remarkable constancy [in voltage and frequency] had arisen'.[15] While the constancy of the frequency stood at the centre of their design, the precise value of the frequency was less important for their measurements. Yet, they soon found another military application for the triode-maintained tuning fork – secret transmission of pictures, which hinged on precise synchronization of frequencies.

A few methods for transmission of pictures by electromagnetic signals were known at the time. Eccles and Jordan based their system on the 1904 method of the German physicist Arthur Korn. As with most facsimile methods, in Korn's method an electric 'eye' moved in front of the picture, scanned it and translated its luminosity into electric signals. A light beam repeated the motion of the 'eye' in the receiver, producing a copy of the picture on a chemical paper. Korn suggested a mechanism by which the motion of the source controlled the motion in the receiver to ensure synchronization. Instead, Eccles and Jordan controlled the period of each end through a separate tuning fork, dispensing with the need to send signals about their motion. Since one could not reproduce the picture without knowledge of the period of the 'eye', the system could be used for secret signalling.[16] The system required high degree of agreement in the frequencies of the two tuning fork circuits. Since the mechanical period of the 'electric eye' was much lower than that of a tuning fork, Eccles and Jordan needed a mechanism to reach lower frequency. This was easy to find; a means to couple the tuning fork to low frequency electric oscillators was well known. In 1875, a Danish inventor Poul la Cour had invented the phonic wheel – a kind of electric motor that turns at a known fraction of a tuning fork's frequency – for synchronization in telegraphy. Rayleigh claimed to have invented the device independently for his needs in acoustical research, three years later.[17] As with Abraham and Bloch, the novelty of Eccles and Jordan was in connecting known mechanic and electric methods to the new triode valve.

Frank E. Smith, the head of the NPL division of electrical standards and measurements, used Eccles and Jordan's triode-maintained tuning fork to measure very short time intervals. Smith did not try to improve telecommunication, but ballistics. In his war project he measured projectiles' velocities. With David Dye he recorded the electric current induced by projectiles passing through coils on running cinematograph paper and compared their marks with steady marks of time. For the latter end Smith 'used first a purely electrical arrangement, and found that a very constant frequency (about 1,000 per second) could be obtained … For certain reasons, however, it was thought desirable to employ a tuning fork, and one of those had been borrowed from Professor Eccles'. The marks made by the tuning fork attained 'an error not greater then one fifty-thousandth of a second'.[18] This was probably the first application of an electronic circuit for precise time measurement, although the tuning fork had already been used for timing before in mechanic and electromagnetic instruments. Smith's transformation of the electronic tuning fork method from frequency to time measurement displays the close connection between standards of time and frequency.

Dye and the NPL

Ballistics was not part of the regular expertise of the NPL's division of electrical standards. Normally, the division engaged with studies related to electromagnetism and its commercial use. With the development of radio communication, frequency measurements became a central concern of the division, answering a growing interest from the government, the military and private companies. With the end of hostilities the military remained the major user of wireless communication with high stakes in measuring and controlling frequencies. High frequency (by contemporary standards), stability and accuracy of senders remained a central concern of the Admiralty, which requested apparatuses for measuring wavelengths from the NPL throughout the 1920s. With the rapid development of civilian wireless broadcasting, the British post office, responsible for civilian communication joined the request for high frequencies standards at the middle of the decade; towards its end a commercial company like Marconi's Wireless Telegraph Co. deemed the field useful enough to fund research carried out at the NPL.[19]

Until 1921, wireless was used overwhelmingly for point-to-point (or to a few points) two-way communication by the military arms, a few commercial companies and amateur operators. The swift emergence of public broadcasting, i.e., transmitting signals from one broadcaster to a large number of listeners who are unable to transmit signals back, posed new challenges for governments. Along with the growth of two-way transmission, the emergence of broadcasting stressed governmental interest in dividing the useful electromagnetic spectrum to many communication channels. To this end governments had to allocate a relatively small band for each station. Interference between transmissions at overlapping, or even nearby, wavelengths posed another problem for the allocation and regulation of the electromagnetic spectrum.[20] Such regulations meant standards in two senses of the term: as specifying the technical requirements, like the deviation of the actual frequency form the allocated one, and as providing the means to check whether these rules are kept. The latter required refinement in the precision and accuracy of measuring methods.

Within the NPL, frequency standards became the expertise of Dye. Dye joined NPL's electrical measurement division in 1910, after studying engineering at the London Guilds Technical College. 'A brilliant but rather irascible scientist', 'he showed ... a wonderful instinct for measurements of the very highest accuracy; and especially for the attainment of this accuracy by means of perfection of the mechanical construction of his instruments'. Thus, upon Smith's departure in 1919, the thirty-two-year-old Dye succeeded him as the head of the division. While heading the division Dye returned to formal studies, attaining a Doctor of Science degree in 1926.[21]

The war-related research provided Dye new techniques for measuring high frequency. In 1919 he adopted Abraham and Bloch's multivibrator and their methods in examining extant standards of radio frequency. Improving on the inventors he combined, probably for the first time, the multivibrator to the triode-maintained tuning fork, on which he had learnt from Eccles and Jordan. Well familiar with the use of the latter from his research on ballistics, Dye saw its combination with the multivibrator as a simple step.[22] He regarded the combined system as the future standard for high frequency and continued improving it in the following years. Dye directed his efforts into two main goals: increasing the precision of the system, and extending its use for higher frequencies. By 1922 he could measure frequencies of 10^7Hz. To this end he connected two multivibrators in cascade.[23] Already in 1921 the NPL was satisfied enough with the accuracy to adopt the tuning fork–multivibrator device (with a one multivibrator) as its standard for radio frequency in its reliable range of up to 150 KHz (below broadcasting range). Yet, as is common in the work on standards, which includes recurrent refinements, Dye continued examining the precision of the apparatus, and especially its most important component – the electronically maintained tuning fork. He experimented with variations in the frequency of a tuning fork under changing physical conditions like temperature, magnetic field and modifications in the triode circuit, comparing the affected tuning fork to one that was 'kept invariable as possible' under constant temperature and pressure.

But how does one know that a tuning fork circuit under invariable conditions actually keeps a constant frequency? To this end it should be compared to a stable standard. Since frequency and time are two sides of the same periodic motion, the tuning fork periods could be compared to those of a standard astronomical clock. Therefore Dye needed a way to count the tuning fork's vibrations. He constructed a more elaborate system than Abraham and Bloch who marked the vibrations of the tuning fork on a cinematic tape and compared them to those of a standard clock. Forty years earlier, the instrument maker Rudolph Koenig designed a mechanical clock controlled by tuning fork. Like Dye, Koenig did not have a direct interest in horology but devised a clock to determine his tuning forks' frequencies by comparing the rate of the clock to that of a standard pendulum clock.[24] Dye followed a similar track, yet he relied on the new triode technology. With that technology, Dye constructed what was arguably the first electronic timekeeper.[25]

Following Eccles and Jordan's design for secret facsimile, Dye employed a twenty-tooth phonic wheel to reduce the 1,000 Hz vibration of the tuning fork. The phonic wheel drove a 50:1 worm wheel, which closed an electric circuit 'once each 1000 alternations', marking thereby a dot on a chronograph tape, in a manner similar to that suggested by Smith to record the motion of projectiles. Moreover, by adjusting the electromagnetic properties of the circuit, the motor that drove the chronograph was put in synchronization with the tuning fork. 'In this way the tuning fork records its own frequency directly on the chronograph without any attention and with extreme accuracy'. Comparing the 'second' dots made by the tuning fork with those marked by a standard second pendulum clock, Dye observed the accumulated error in the period of the former. With assistants he continued to carry such comparisons for longer intervals of up to a week on the same tape. By 1932 '[t]he frequency stability over hourly periods [was] of the order of 5 parts in 10^8, and over weekly periods, 3 parts in 10^7'; the latter was at the same order of magnitude as the best mechanical pendulum clocks of the time, and the short-range accuracy approached that of the most exact electro-pendulum clocks.[26]

Beyond the purposes of wireless communication, for which he had begun the research, Dye regarded the tuning fork circuit as a precision time standard. As such he deemed it useful for measuring relatively short intervals of time (as he had done with Smith), possibly even 'to observe variations in the hourly rate of standard clocks'.[27] He had not, however, connected the tuning-fork-controlled phonic wheel to a clock mechanism that would allow continuous reckoning of time and its continuous display. That Dye did not take this step does not seem to originate in technical difficulties. It rather reveals his lack of interest in making such a continuous display tuning fork clock, as it did not seem to serve any concrete aim, and its construction did require meticulous work.

Bell's Tuning Fork Clock

While state agencies sought exact standards of frequency for coordinating wireless communication under their jurisdiction, AT&T needed them to integrate its extended telecommunication system, known as the 'Bell system'. During the 1910s the world's largest telecommunication company introduced electronic technologies to its network. These included new radio techniques, which allowed transmission of the human voice over a distance, to complement its extensive wired network and the array of services it provided. It also developed new methods to multiply the number of conversations that could be transmitted over its wired network (called multiplex telephony). A common standard for measuring these oscillations was a prerequisite for coordinating the expanding number of methods in use, as it was necessary for attaining smooth transmission of signals between them. Accurate standard allowed also increasing the number of transmission in multiplex telephony, and in the wireless waveband allocated to the corporation, as each could be confined to particular frequencies. Its construction was a task given to a group of researchers with backgrounds in physics and engineering headed by Joseph Warren Horton, a physics graduate, at the corporation's research department.

For the basic frequency standard of the Bell system, Horton's team adopted the French and British novel triode maintained tuning fork technologies, albeit in a modified arrangement suggested by researchers of the American Bureau of Standards (BoS). Since the Bell system required precise knowledge of a wide range of frequencies, Horton's group invested much effort in refining the stability and accuracy of their standard, and in methods for multiplying its value. To allow sensitive measuring of frequencies to within 100 Hz at a wide range, the group chose a tuning fork of that period (ten times lower than Dye's), and devised a special system for reaching any multiplication of its value up to 100,000 Hz. Still, the accuracy of the system clearly depended on that of fundamental tuning fork. Within the tradition of exact measurements in acoustic and independently from Dye, Horton suggested comparing the tuning fork, which should have 'the general characteristics of a good clock', with an astronomical timekeeper. Unlike Dye, Horton was not satisfied with connecting the tuning fork only to a tape chronograph; instead he suggested controlling a clock mechanism by the tuning fork.[28] By June 1923 Bell's tuning fork drove a clock mechanism through a synchronous motor and a commutator that reduced the frequency of Horton's design. Unlike the chronograph, the tuning fork clock moved continuously, allowing the observation of small errors as they accumulated over time. The same mechanism that transferred the tuning fork vibration into a mechanical rotation for driving the clock allowed marking of time signals on a chronograph for monitoring possible fluctuations over shorter intervals. The researchers invested much effort in developing a suitable synchronous motor for driving the continuous clock mechanism.[29]

Arguably, the coupling of the basic frequency standard to a clock was the greatest and most important novelty of the system. The idea and the means to accomplish it had precedents. Still, both the linkage of radio frequency to continuous time measurement and the construction of a clock on an electronically maintained tuning fork (which suggests a steadier operation than Guillet's earlier microphone mechanism) were important original steps of the group. Dye followed a similar track but was satisfied in comparing the NPL's frequency standard to a timekeeper over limited intervals. Horton's group, however, conceived AT&T's frequency as a continuous reference for tuning and measuring frequencies of electric oscillations at the Bell system. The NPL, on the other hand, regarded its central standard as a means for calibrating other devices used as secondary standards; its continuous operation was not deemed important for its main purpose. Since Bell's frequency standard operated continuously, its researchers sought a continuous method to inspect its performance.[30]

By comparing the tuning fork clock to the laboratory clock, the group concluded that the former is exact to within 'about 6 parts in 1,000,000', well within the needs of the Bell system, set in 1923 at one part in 100,000. Still, the group continued improving the accuracy of its standard, reaching an accuracy of three parts in 1,000,000. Since this accuracy was higher than that of their laboratory electric pendulum clock, the researchers established it by direct comparison with the radio signals from the naval astronomical clock. The accuracy of the tuning fork clock was only one order of magnitude lower than that of the BoS's standard clock. Still in 1927, Horton's group suggested improvements in the mechanism that would increase the accuracy by about twenty-fold; as mentioned, such accuracy was attained later by the NPL in its chronometer. The increase interest of the Bell group in horology can be seen in its use of the clock fork for exact time measurements. In January 1925, Warren Marrison at the laboratory recorded the timing of the total solar eclipse at different locations in the northeast USA. Stations at each of these locations sent a signal through telephone line to the laboratory in New York, where it was marked against the time signals from the tuning fork.[31]

In autumn 1927, when Horton and Marrison announced the higher new accuracy of their tuning fork clock, it had already been eclipsed by their own new quartz clock. Soon, workers in the field adopted quartz frequency standards, as they proved more stable and could reach wider range of frequencies that became useful for electric communication than the tuning forks. Although based on different physical phenomenon – piezoelectricity, the mutual influence of electricity on pressure in crystals, the research on the new quartz standards continued that on the tuning fork. Many of the electronic techniques in use were very similar. In particular, Marrison followed the work on the tuning fork standard in developing a quartz clock to monitor the period of the piezoelectric frequency standard. With their high accuracy, quartz clocks became the most popular device for exact timekeeping, replacing the pendulum clocks, and apparently obstructing the development of tuning fork clocks. Tuning fork controllers became important again with the minimization of electronics after the advent of the transistor. An electronic tuning fork watch developed in the late 1950s paved the way for the quartz watch, which has since became ubiquitous, as the electronic tuning fork clock paved the way for the quartz clock in the 1920s.[32]

Conclusions

The triode-maintained tuning fork clock, the first electronic clock, resulted from an accumulation of small steps. Innovations consisted of quite minor additions and modifications of previous methods, which in retrospect often seem straightforward or even trivial. Once the triode became a central powerful device of wireless technology, connecting the hitherto electromagnetically maintained tuning fork to the electronic valve was quite straightforward; although Abraham and Bloch did not connect the new device to their multivibrator, to Dye such a connection seemed quite trivial. Since the idea of comparing tuning fork frequency to a clock was well known, it seems simple to connect its electronically maintained version to a chronograph; the means to do so were known and only required some modifications: the phonic wheel motor, the chronometer and the mechanism for marking dots on tape. Driving a clock dial, and not only a dot-marking mechanism from the phonic motor, did not require much imagination, especially as the phonic wheel had already been used for timers. Arguably the only innovation that was not based on a previous device was that of the multivibrator, which still relied on a known idea of producing harmonics. This gradual process does not offer great 'Eureka' moments, yet it represents many important inventions.[33] And, indeed, it was a process of invention. It was not merely a process of refinement and improvement, as the end result – a highly accurate electronic clock and a system of radio frequency standards – is clearly different from the original electromagnetic tuning fork. Moreover, this electronic device opened the way for overthrowing centuries-old mechanical horology by the quartz clock.

Still, carrying out these small steps required specific reasons for contriving the innovations, knowledge of and preferably experience with the related technologies, and some ingenuity. Otherwise many more researchers would have suggested each innovation. To take an example: that two resourceful researchers like Abraham and Bloch did not connect their own multivibrator and triode-maintained tuning fork suggests that the step required some ingenuity. That Guillet did not incorporated triodes in his new electromagnetic tuning fork clock suggests the crucial role of the inventors' experience with electronic technologies, in addition to their knowledge of the tuning fork methods, of which also Guillet was an expert. The researchers discussed in this article designed their methods for specific aims, like measuring amplification power of triodes, synchronization in secret signalling, measuring velocities of projectiles, or the frequencies of different radio waves. The crucial role of their specific aims is well illustrated by the difference between the otherwise equivalent uses of the tuning fork frequency standard at the Bell system and at the NPL. Since only Bell required continuous reference to measure frequencies in its system, Horton's group constructed the first electronic clock, while Dye was satisfied with a chronometer.

Most, but not all, of these specific technological goals originated in efforts of improving methods of wireless communication for military, commercial and governmental interests. Under the pressure of the Great War, scientists and engineers applied their expertise in exact measurements and in electronics for military aims, resulting in triode tuning fork methods. Nevertheless, Dye, Horton and their colleagues developed the electronic tuning fork frequency standards for the novel needs of telecommunication as a mass technology. Precise knowledge of frequency was necessary for an efficient coordination and integration of large telecommunication system, providing an immediate reason to construct accurate frequency standards. Exact frequency standards answered the commercial interest of AT&T and the social interest of the government in regulating and fostering radio communication. While standardization had a social function, its roots were in scientific practice. Scientists and scientific instrument makers had turned the tuning fork into a high precision instrument for laboratory measurements of acoustics and time. Scientists and academic engineers transferred the tuning fork from a research instrument to a central device of wide scale telecommunication, which required precision hitherto limited to the scientific laboratory. While Horton, Dye and their colleagues developed basic tuning fork standards for the techno-social goals of their institutes, like good metrologists they sought to improve the precision of their standards, also beyond their practical goals. That the major use of the tuning fork clock qua clock was to help investigating natural phenomenon (sun eclipse) suggests that the precise instrument still had a scientific value, even when it served the technological need of a giant corporation.

9 CALIBRATING THE UNIVERSE: THE BEGINNING AND END OF THE HUBBLE WARS

Genco Guralp

Introduction

Historians of science generally agree that the linear relationship that Edwin Hubble discovered, between the velocities of galaxies and their distances, constitutes a major breakthrough in the history of observational cosmology.[1] In addition to putting the *expanding universe* paradigm on a firm footing, Hubble's work provided the impetus for a research program that still continues to be of central importance in experimental cosmology today.[2] This is mainly due to a key component of the velocity–distance relationship, namely, the *rate of expansion of the universe*, known as the *Hubble constant*. The measurement of this constant is considerably difficult and depends crucially on correctly calibrating different methods of astronomical distance determination. As a matter of fact, Hubble's own value of the constant suffered heavily from calibration problems and was stricken by gross systematic errors. He was also aware that the result was obtained for a small distance scale and had to be confirmed for larger distances as well. Consequently, he 'initiated an exploratory program to follow the relationship to the greatest distances attainable with the largest telescope'.[3]

The exploratory program that Hubble initiated was brought to fruition by his student, Alan Sandage, one of the most influential astronomers of the twentieth century. Through his observational program, Sandage set the stage for the post-war attempts to determine the value of the Hubble constant. His calibration scheme of 'precision indicators' consisted of selecting the best standard candle[4] at each level of the cosmic distance scale and following this singular technique of calibration throughout the measurement process.[5] This approach was opposed by the French astronomer, Gérard de Vaucouleurs, who meticulously devised an alternative calibration scheme of 'spreading the risks', which advocates the methodology of using as many techniques as possible and then averaging over them. For more than three decades, these towering figures produced incompatible values for the Hubble constant, based on their respective methodologies. Sandage obtained a value of 50 km/s/Mpc, whereas de Vaucouleurs insisted that the correct value was 100 km/s/Mpc.[6] This conundrum, which then came to be known among astronomers as the *Hubble Wars*, was only resolved in the early 2000s, mainly due to the efforts of a collaboration known as the *Hubble Space Telescope Key Project*. This collaboration was specifically formed with the aim of determining the Hubble constant up to an accuracy of ±10 per cent.

In this paper, I offer an account of the emergence, development and the resolution of this measurement conundrum. I analyse the history of the efforts to measure the correct value of the Hubble constant, on the basis of a theoretical framework that Hasok Chang introduced in his book *Inventing Temperature*.[7] I argue that even though each stage of this historical episode exemplifies Chang's notion of *iterative progress*, this notion by itself is not sufficient to understand how the conundrum was resolved. For a better understanding of how the Hubble Wars ended, I claim, we need to situate it within the general transformation that cosmology underwent in the early '90s, which is known as the *precision era*. More specifically, drawing on Galison's work,[8] I urge that within the precision era, a new *material culture of calibration* came into play in experimental cosmology, in which various working groups using different methods sought error reduction through precision measurements as their primary goal, as opposed to a 'philosophical' commitment to a single methodology that we see in the cases of Sandage and de Vaucouleurs.[9]

The plan of the paper is as follows. First, I outline Alan Sandage's observational program and his methodology for measuring the Hubble constant. Next, I discuss de Vaucouleurs's approach to the problem and his attack on Sandage's methodology. I then describe the work of the Hubble Key Project and examine how the results of this project ended the controversy over the value of the Hubble constant. Finally, I offer a critical analysis of this episode and argue that a synthetic reading of Chang and Galison's works can help us to understand the mode of scientific progress that this case presents.

The Sandage Programme

In an article he published in 1970 in the journal *Physics Today*, Allan Sandage famously characterized cosmology as a 'search for two numbers'.[10] This was seven years before the *Hubble Wars* was officially launched by Gérard de Vaucouleurs, with his publication of the first paper of his eight-article series on the *extragalactic distance scale*.[11]

The importance of Sandage's *two-numbers* article, as it then came to be known, stems from the fact that it provides a very clear statement of his programme for observational cosmology, and in particular, for the determination of the Hubble constant. Sandage carried out the observational programme outlined in the article during the decade following its publication. In the present section, I first provide the background to Sandage's research before embarking on his long-term programme to determine the Hubble constant, by focusing on a cornerstone paper that he co-authored with Nicholas Mayall and Milton Humason. Occupying 'nearly an entire issue of the *Astronomical Journal*', as one commentator notes, this paper was the result of a 'multiyear effort to determine the nature of the expansion of the universe'.[12] After analysing the main conclusions of this paper, I examine the *two-numbers* article to outline the elements of Sandage's programme for experimental cosmology. I then study the series of articles he published between 1974 and 1981 under the general title 'Steps Towards the Hubble Constant', in which he laid out and executed his strategy of 'precision indicators', for the measurement of the Hubble constant.

Before I begin unpacking the details of Sandage's programme, I would like to say a few words about the context of the search for the correct value of the Hubble constant. Firstly, one should note that the post-war period cosmology was dominated by the historic debate between the steady state and the big bang models. This question was only to be resolved in a definitive manner (in favour of the latter) with the advent of the measurements of the cosmic microwave background radiation, first detected by Penzias and Wilson in 1964.[13] Secondly, many cosmologists of the era shared the conviction that the expansion of the universe was decelerating. Finally, it was believed that the cosmological constant was zero.

All these factors contributed to the *measurement value* of the Hubble constant in experimental cosmology. As a key element for understanding the expansion history of the universe, a precise knowledge of the constant, as Sandage argued later on, could help select the correct model of the universe, and in particular, determine the 'age of the universe', quite straightforwardly.

The HMS Catalogue and the Completion of the Hubble Programme

In 1956, Allan Sandage, with his two colleagues, Mayall and Humason, published a sixty-five-page long article entitled 'Redshifts and Magnitudes of Extragalactic Nebulae', known as the *HMS catalogue* among practising astronomers, after the initials of its authors. The importance of this article, which contained results of observations made 'during the 20-year interval from 1935 to 1955',[14] stems from the fact that it represents the culmination of the research programme that Hubble started in 1929.[15] With data from over 800 nebulae, the paper represents the largest survey of its time. It is divided into three major sections, separately written by each author. The first two sections, composed by Humason and Mayall respectively, present the data and the third section by Sandage provides the analysis.[16] It was in this final section that Sandage pointed out an important mistake in Hubble's calibration of the distance ladder and announced the revised value of the Hubble constant.

In particular, Sandage targeted two questions in the analytical part: on the one hand, he inquired into whether the data was reliable, and on the other, to the extent that this was the case, he examined the empirical question of whether the velocity–distance relationship remained linear in large distances as well. He dealt with the question of the numerical value of the Hubble constant only in an appendix and in a very tentative manner. At the very beginning of his discussion of the constant, he indicated the difficulty of the task at hand as follows:

> The determination of the expansion parameter H is one of the most difficult problems in modern observational astronomy, since each step required for an accurate solution is just on the borderline of possibility.[17]

Sandage then noted that the main difficulty consists in finding a solution to the following dilemma: one has to observe objects distant enough so that the motions unique to individual galaxies are overcome by the general expansion. Yet, the reliability of distance indicators decrease the further out one goes, so objects have to be close enough for the measurements to be accurate. A common strategy to get out of this dilemma is to identify various astronomical objects within distant galaxies that can be used as 'standard candles', whose intrinsic brightnesses can be determined by relatively better known objects such as Cepheid stars. If one can discover these objects at larger distances, one can then use their apparent brightnesses to infer their distance, using the inverse square law.

The key contribution of the paper concerning the Hubble constant was showing that Hubble's attempt to use this strategy was erroneous, which meant a reversal of the entire scheme of calibration that Hubble employed. As Sandage went on to explain, Hubble's calibration depended on the correct identification of the 'brightest resolved objects in a sample of nearby resolved nebulae'.[18] The particular calibration scheme Hubble used was as follows: the brightest objects in a given nebula were used as standard candles. Hubble believed them to be 'supergiant stars'. The absolute magnitudes of these objects were assumed to be known from the cepheid calibration of the blue supergiant stars M31 and M33. Lastly, the precise value of Cepheid brightnesses was determined from parallax observations, which constitute the lowest step of the ladder.

The results of the HMS catalogue showed that all these assumptions and determinations were questionable. First of all, starting from 1945, it was noticed that the zero-point of the period–luminosity relation that Hubble used in his calculations was erroneous. This revision in the Cepheid distance already affected all the higher steps of the ladder. In addition, Sandage realized that the key assumption that Hubble made in his calibration was mistaken: the brightest objects observed by Hubble were not stars but H II regions.[19] Sandage managed to identify stars within these regions but they were much fainter than Hubble assumed them to be. As he put it:

> ... although it will be possible to use the brightest resolved stars as distance indicators, they are faint and must first be isolated from the H II regions.[20]

This sentence gives the key to Sandage's attack of the problem of Hubble constant for the years to come. Once the stars are identified and separated from the H II regions, a re- calibration routine would be followed. Still, even within the limits of the HMS catalogue, Sandage believed one can give a provisional value for the constant. For his readjustment of the expansion parameter, he offered two arguments as two 'ways' of obtaining its value:

Use of Andromeda Nebulae: The rich dataset the HMS contained gave Sandage enough confidence to assume that the nebulae had an upper luminosity limit. For calibration purposes, he further assumed that the Andromeda nebula is intrinsically the brightest object among the nebulae. Admitting the arbitrariness of the assumption, he justified it by the fact that the value of the constant thereby obtained coheres with the one that obtained from the second method.

Use of Stars in NGC 4321: As a second method of calibration, Sandage used the apparent magnitudes of the brightest resolved stars in the NGC 4321 galaxy. By comparing with the known values of the brightest stars in M31 and M33, one can obtain the absolute magnitude information. Combined with the redshift value of the Virgo cluster, to which the galaxy belongs, one can obtain a value for the Hubble constant.

Using these methods, Sandage obtained the value of 180, although he was careful in his statement of the result: 'Although it is probably uncertain by 20 per cent, $H = 180$ km/sec 10^6 pc ... appears to be the best obtainable from the present data'.[21] It was clear to Sandage that a new observational programme was needed.

The Two-Numbers Article and the Beginning of the Sandage Programme

Whereas the HMS catalogue was written in the full spirit of the observational programme of Hubble, the two-numbers article represents a new context of observation which differs significantly from the previous one. Sandage now articulates the measurement of the Hubble constant together with the deceleration parameter q_0 'as a crucial test for cosmological models'. Thus whereas in the HMS catalogue, the Hubble constant made only a tentative appearance in an appendix, it is now the main target of research, with a key role to play in adjudicating between different world models. Yet, Sandage again warned his readers concerning the difficulty of the enterprise:

Although the observer's problem, to find H_0 and q_0, is easy to state, it has defied solution for 40 years.[22]

After this pessimistic observation, Sandage immediately introduced the main problem which would make its mark on the next episode of experimental research: *distance calibration*. He put the problem as follows:

> Distance calibration is a stepwise procedure, with the errors proliferating with each step. First one measures the apparent brightness of certain well defined objects, the distance indicators, in the nearby resolved galaxies. If the absolute brightness of these indicators is known from a reliable previous calibration, the distance follows from the inverse-square intensity fall of. Because a unique relation exists between the period and absolute luminosities of Cepheid variable stars these stars are excellent distance indicators.[23]

It is important to note here that when Sandage uses the term 'calibration', he does not refer to a particular measuring device. Rather, what is meant is the correct conversion of an astrophysical object's apparent brightness into an empirically valid distance information.[24]

The central theme of the story is contained in the first sentence in the above quote: *calibration is a stepwise procedure*. As the measurement of the Hubble constant fundamentally depends on measuring distances, it is the determination of the distances that involves this stepwise procedure. It is no coincidence that the central series of papers on the measurement of the Hubble constant that Sandage produced throughout his career carry the title: 'Steps towards the Hubble Constant'. With his long-time collaborator, the Swiss astronomer Gustav Andreas Tammann, he wrote ten papers, which span a time frame of twenty-one years. In these papers, they determined each step of the distance ladder, in which they executed (almost word by word) the plan that was introduced by Sandage at the end of the HMS paper and elaborated further in the 1970 *two-numbers* article. Below, after pointing out several key points that Sandage made in this article, I will focus on the *steps papers* in the next section.

The stepwise procedure, as we will see below, requires many methodological decisions to be made. But in order to understand how these decisions are made, their *material context* needs to be taken into account. As Sandage explains, the resolution capacities of telescopes play a role in the very definition of the range of measurement. In other words, as opposed to cases in which the measured quantity is within the range of the device capacity so that the measurement range is determined by theoretical considerations only, one sees in this case that the very definition of the measurement range depends on the device capacity and hence changes as the telescopic resolution gets better:

> The crucial distance range within which H_0 can be determined is quite narrow. It extends between 10^7 light years, which is remote enough so that expansion velocities

begin to dominate the spurious velocity effects and 6 x 10^7 light years, which is the *upper limit for the indicators to be resolved in nearby galaxies with the 200-inch Hale telescope.* In this range, indicator objects include the brightest resolved red and blue supergiants, the angular size of H II regions, normal novae, and perhaps, after much new calibration, supernovae. *Each of these classes must first be calibrated in even nearer galaxies, less than 10^7 light years away, where the more precise distance indicators of Cepheid variables can be measured.*[25]

The distance range has to have a lower limit because in shorter distances the peculiar velocities of the galaxies dominate the expansion velocity and this 'masks' the systematic effect of the expansion. The importance of the 'nearby' field is for calibrating in a precise way the standard candles to be used for further out distances. This latter calibration forms the backbone of the research programme that Sandage followed with the *Steps* articles.

The *two-numbers* article ends its discussion of the Hubble constant by reporting on the corrections made to the local distances and on the basis of these corrections it extrapolates the value of the constant to $15 \leq H \leq 40$. This almost eighty percent fluctuation in the value of the constant, which was tentatively given as 180 in the HMS catalogue, is one dramatic illustration of the difficulty of measuring it.

The Steps Series

The first paper of the *Steps* series, which was published in 1974, is titled: 'Calibration of the Linear Sizes of Extragalactic H II Regions', in line with the programme outlined above. The opening of the paper neatly lays out the fundamental problematic as follows, in a way which is very similar to the earlier programmatic article:

> ... the Hubble rate is extraordinarily difficult to measure directly because distances must be determined with high precision to galaxies that are so remote as to have significant expansion velocities. Cepheids have long since faded below plate limit for such galaxies. The redshifts must be large enough so that the effect of mean random virial motions, or of any local velocity perturbation, can be neglected. This requires new calibration of precision distance indicators that are brighter than Cepheids and that enable us to reach these distances. *Much of the work discussed in this series of papers concerns the isolation and calibration of such indicators.*[26]

This problematic of reaching higher distances sets the stage for the entire debate between Sandage and de Vaucouleurs, which could perhaps be more aptly called the *calibration wars*. For once the problem of reaching the higher steps of the distance ladder is identified, the question of method, that is, of *how to proceed* is to be addressed. Here is how the *calibration recipe* given in the first Steps article:

The linear sizes of H II regions in late-type spirals ... are calibrated using galaxies whose distances are known from Cepheids...

This size calibration is extended to include supergiant spirals. The distance to the nearest of these (M101) is found in Paper III using six methods, including use of the brightest stars as calibrated in Paper II.

The H II region sizes are used to determine distances to 50 late-type field galaxies in the distance interval $m - M < 32$ (Paper IV). The *distribution of absolute magnitudes* of the galaxies in this 50-galaxy sample, as a function of luminosity class (Paper IV), follows from these data.

Redshifts of newly identified giant Sc I spirals with $m - M > 35$ have been measured as the last step. Combining the redshifts and the absolute magnitudes of step 3 gives H_0.[27]

In Step (1), one aims to reach a correct estimate of the diameter of the H II regions using cepheids. In order to do this, *galaxies as cosmological objects* are used as *calibrators* through a *physical property* roughly as follows: it is assumed that the size of the H II regions correlate with the brightness of galaxies that they belong to. Thus, if one can obtain the distance to the galaxy from an independent method, the H II region size will be calibrated and the ladder can be extended further. Step (2) is a variation of the first one, again aimed at using the H II regions to give distance information concerning supergiant spirals, which are further away on the distance ladder.

The de Vaucouleurs Objection

In a series of papers paralleling Sandage's *Steps* programme, the French astronomer Gérard de Vaucouleurs offered a staunch criticism of Sandage's scheme of calibration and attempted to replace it with his own alternative. The main point of attack was that Sandage's method, for each step of the ladder, relied on a single procedure which carried enormous weight for the structure that is built on it. In the first paper of the series, we read the following 'manifesto' by de Vaucouleurs:

> Tradition notwithstanding, distances derived from Cepheids calibrated in open clusters deserve no greater weight than the others. The unending discussions, revisions, and rediscussions of the P-L, P-L-C, P-L-A[28] relations make the point clear. Because of possible effects of age and chemical evolution on any indicator, it is risky to rely primarily or exclusively on any one indicator, while it is unlikely that all are affected in the same sense and amount by evolutionary differences between galaxies. Rather than rely entirely on a select few so-called 'precision-indicators', the basic philosophy of 'spreading the risks' will be adopted here.[29]

In the eighth paper of the series of papers he wrote, executing his 'philosophy' of 'spreading the risks', de Vaucouleurs obtained the value of 100 ±10. This contrasts with Sandage's results sharply. Even though Sandage announced many results throughout his career, from the beginning of the research programme of the *Steps* series, all his values centred around 50. For example, in the sixth paper of the *Steps*, published in 1975, Sandage and Tammann obtained: 56:9 ± 3.4.[30]

In a 1982 monograph entitled *The Cosmic Distance Scale and the Hubble Constant*, de Vaucouleurs gave a more general and comprehensive characterization of how the two 'approaches' – as he called them – differed, as follows:

> In the treatment of the galactic extinction corrections: whereas the Sandage approach assumes a relatively low galactic extinction, de Vaucouleurs assumes a significant value for this.
>
> In the choice of primary indicators: Sandage uses only one fundamental indicator (the cepheid period–luminosity–colour relation) – *putting all their money on one horse, as it were* – while the author uses no less than five ... following a philosophy of 'spreading the risk'.
>
> In the number of calibration methods: Sandage and Tammann use again only one technique – thus doubling their bet, so to speak, while the author used no less than *ten* methods (of which nine are independent...) to fix the zero points of his five independent primary indicators.
>
> In the number of secondary indicators: Sandage and Tammann uses three of them, two of which depend for their calibrations on precarious extrapolations. The author used *six indicators* ... and carefully avoided any extrapolation of the calibrating relations.[31]

Although de Vaucouleurs lists a couple of other discrepancies, the above list contains the core ones and all of the items (1) to (5), in one way or another, relate to the question of calibration. For, in order to calculate the Hubble constant, two fundamental steps of calibration are required. Firstly, the distances to the nearest galaxies are evaluated by means of primary stellar distance indicators that are *calibrated by fundamental geometric or photometric methods in our Galaxy*. Secondly, a scale of relative distances to more distant galaxies is constructed by means of secondary and tertiary distance indicators calibrated in the nearby galaxies. As both these steps depend crucially on the galaxy extinction model that is used, it also forms a part of the calibration process.

For de Vaucouleurs, this situation presented a calibration conundrum that can be resolved on the basis of the following *methodological* principles:

(i) use the largest possible number of different distance indicators and *independent* methods of calibration in order to minimize accidental and systematic errors;

(ii) avoid extrapolations and use calibrating regressions only as *interpolation* formulae for reduction to central or mean values of the parameters;

(iii) beware of circular reasoning, unverified assumptions, and subjective choices based on intuition, predilection or plausibility arguments;

(iv) subject the distance scale so constructed to a multiplicity of *a posteriori* checks ... by means of *independent* distance indicators not previously used in its construction.[32]

Armed with these methodological principles, de Vaucouleurs applied his philosophy of *spreading the risks* meticulously: he had ten different calibration methods. He wrote with authority:

> To construct a distance scale on a secure basis a philosophy of 'spreading the risks' must be adopted. The belief that a single indicator (such as the cepheids) is superior to all others has no factual basis as a realistic assessment of errors indicates (emphasis added).[33]

As a result of all this, de Vaucouleurs came up with the value of 100, which is double the value Sandage promoted.[34]

How did Sandage respond to these criticisms? His main line of counter-attack was accusing de Vaucouleurs of falling prey to the *Malmquist* or *selection-bias*. Basically, the *Malmquist bias* is the usual selection effect which stems from the fact that as one reaches higher and higher distances, the data points one selects will be intrinsically brighter objects, which will introduce a bias into one's data. As expected, de Vaucouleurs emphatically rejected this criticism, claiming that it was based on a misunderstanding. Referring to Sandage and his co-workers, he wrote that the Malmquist bias is:

> elaborately discussed by Sandage, Tammann & Yahil ... and freely invoked by them to discredit the work of others when it is in disagreement with their own conclusions concerning the extragalactic distance scale and the Hubble constant.[35]

The personal tone of these lines attests to the fact that the debate between the two scientists assumed the form of hostility towards the end. The issue of Hubble Wars, as the astronomical community chose to call it, was not to be resolved by either of them.

The Concordance Value

Ironically, neither of the values promoted by Sandage and de Vaucouleurs are regarded as correct today. For example, the recently published paper by the WMAP mission, which did not directly measure the Hubble constant but derived it from extremely precise measurements of the cosmic microwave radiation, announced the value: $69:32 \pm 0:80$.[36] A more recent measurement of the Hubble constant, by the *Planck Mission* of the European Space Agency, produced the value of

$$67:80 \pm 0:77,$$

which represents the most precise value as of today.[37]

But the critical work that practically ended the controversy was the *Hubble Key Project* (HKP), which began in the mid-1980s. Here, without going into the details, I want to quote from their penultimate paper to introduce several pivotal points that show the contrast with the Sandage–de Vaucouleurs debate.

Firstly, HKP works with an open 'archival data-base'. They write:

> As part of our original time allocation request for the Key Project, we proposed to provide all of our data in an archive that would be accessible to the general astronomical community. We envisaged that the Cepheid distances obtained as part of the Key Project would provide a data-base useful for the calibration of many secondary methods, including those that might be developed in the future.[38]

This is an instance of what I called the 'material culture of calibration', following Galison: the database is open to the community of researchers for purposes of calibrating secondary distance indicators. This communitarian aspect is also indicated in that HKP worked in a way that resembles collaborations in particle physics as opposed to the solitary or very limited collaborative work of Sandage and de Vaucouleurs. In contrast with the single or two-authored papers of the pre-precision era, the HKP paper has fifteen authors, who all worked in different institutions.

Secondly, the pre-calibration data reduction was carried out by a 'double-blind' technique, explained as follows:

> As a means of guarding against systematic errors specifically in the data-reduction phase, each galaxy within the Key Project was analysed by two independent groups within the team, each using different software packages: DoPHOT ... and ALLFRAME ... The latter software was developed specifically for the optimal analysis of datasets like those of the Key Project, consisting of large numbers of observations of a single target field. Only at the end of the data-reduction process (including the Cepheid selection and distance determinations) were the two groups' results intercompared. This 'double-blind' procedure proved extremely valuable.[39]

This form of bias reduction was not a possibility for the solitary authors of the *Hubble Wars*.

Thirdly, the discussion of calibration in the HKP paper differs significantly from the pre-precision era. In contrast to the 'philosophical' debates over the methodology of data analysis, the problem of calibration is dealt with in the material context of the Wide Field and Planetary Camera 2 of the space telescope:

> Ultimately, the uncertainty in the Hubble constant from this effort rests directly on the accuracy of the Cepheid magnitudes themselves, and hence systematically on the CCD zero-point calibration. In view of the importance of this issue for the Key Project, we undertook our own programme to provide an independent calibration of both the WF/PC and WFPC2 zero points, complementary to the efforts of the teams who built these instruments and the Space Telescope Science Institute.[40]

Lastly, HKP presents us with a methodological compromise between Sandage and de Vaucouleurs in their measurement programme. For even though they write that 'The Cepheid period–luminosity relation remains the most important of the primary distance indicators for nearby galaxies'[41] in line with Sandage, they seem to accept de Vaucouleurs' 'wisdom' when they concede that 'Cepheids alone cannot be observed at sufficient distances to determine H_0 directly, and an accurate determination of H_0 requires an extension to other methods'.[42]

Accordingly, the HKP employs four secondary methods, all based on the Cepheid calibration as the primary distance indicator.

The Hubble Key Project was completed in 1999. With the publication of their final paper in 2001, the 'war' concerning the value of the Hubble constant came to an end.[43] As a co-leader of the group, Robert Kennicutt, put it: 'The factor of two controversy is over'.[44]

Analysis

We can analyse this episode starting from the model of scientific progress Hasok Chang offered in *Inventing Temperature*. In particular, Chang's notion of *epistemic iteration* captures remarkably well the methodology of distance calibration in cosmology. However, as I argue below, *epistemic iteration* cannot explain the transition to the *concordance* value of the Hubble constant. A more comprehensive understanding of the resolution of the conundrum can be achieved if we supplement the notion of *epistemic iteration* with Galison's concept of *material culture* in science. Below, I will first introduce Chang's picture, and then explain how it can be combined with Galison's account to analyse the Hubble wars.

Progress through Iteration

On the basis of his case study of the development of thermometry, Chang argues that the circularity involved in empiricist methodology rules out a foundationalist approach to justification in science. He claims that this leaves us with no option but to follow a coherentist strategy. His particular version of this strategy, which he refers to as *progressive coherentism*, is built on the idea that 'the real potential of coherentism can be seen only when we take it as a philosophy of progress, rather than justification'.[45] In other words, Chang thinks that one give an account of scientific progress within the framework of coherentism. The distinguishing character of coherentism, according to Chang, is that a coherentist 'inquiry must proceed on the basis of an affirmation of some existing system of knowledge'.[46] The scientists may have reasons to be suspicious with this system of knowledge, but they still choose to work within its premises for 'they recognize that it embodies considerable achievement that may be very difficult to match if one starts from another basis'.[47] Chang gives the name of *epistemic iteration* to this methodology of critical affirmation:

> Epistemic iteration is a process in which successive stages of knowledge, each building on the preceding one, are created in order to enhance the achievement of certain

epistemic goals. ... In each step, the later stage is based on the earlier stage, but cannot be deduced from it in any straightforward sense. Each link is based on the principle of respect and the imperative of progress, and the whole chain exhibits innovative progress within a continuous tradition.[48]

If successful, this self-corrective process leads to progress, though there is no guarantee that it will succeed. But what counts as progress? Chang thinks that an episode results in progress to the extent that it improves or enhances epistemic values such as simplicity, testability or unifying power. In particular, he distinguishes between 'two modes of progress enabled by iteration': namely, *enrichment* and *self-correction*.[49] As the name suggests, in enrichment, the knowledge system that one works with is not negated but enhanced in various ways, e.g., made more precise or unified. On the other hand, self-correction occurs when the inquiry based on the system leads into its alteration.

I claim that both Sandage and de Vaucouleurs are making use of *epistemic iteration* as part of their methodology. Moreover, the transition from the *two-factor debate* to the *precision era* value of the Hubble constant also involves modes of iterative progress. In general, any cosmological experiment based on distance measurements has to exemplify *epistemic iteration*, due to the very construction of the cosmological distance ladder. There are two main reasons why. Firstly, just as Chang describes, each step of the cosmological distance ladder has to be presupposed to calibrate the next step. Secondly, when the target distance is reached and the Hubble constant is determined, the resulting value has to be checked to see whether it coheres with other parts of our knowledge of cosmology. The case in point is the fact that the age of the universe that one obtains from the Hubble constant cannot be less than the age of the earth or of the oldest stars.

However, I think that the final resolution of the Hubble Wars cannot be understood solely within the purview of iteration. One needs to integrate the historical context into the picture.

To this end, I propose to make use of the notion of *material culture* that Galison introduces in his *Image and Logic*. Galison borrows this term from the anthropology and archaeology literature to capture not only 'the study of objects taken by themselves' but also 'the analysis of objects along with their uses and symbolic significance'.[50] The study of material culture aims at an analysis of objects as 'encultured' or 'entangled'.[51] In this vein, Galison talks about the 'the working physicist's material culture and experimental practices' that 'circulates around the detector. These might include the tools on the bench, the methods of calculation, and the roles of technicians, engineers, colleagues, and students'.[52] One should note that both aspects of the concept of *material culture* are equally important. On the one hand, the concept of *materiality* pertains to the role of detection devices and other technical equipment in knowledge production. On the other hand, the concept of *culture* implies the epistemological decisions[53] that are made by various practitioners on the basis of the technical equipment that are available to them.

Even though I have not fully delineated the forms of *epistemic iteration* at work in Sandage and de Vaucouleurs on the one hand and the Hubble Key Project on the other, certain significant distinctions can still be discerned concerning how the iterative processes are realized. In particular, with HKP, we see that methodological rules are not sacred and compromises can be made for accuracy. Also, the issue of *calibration*, perhaps the key ingredient of epistemic iteration in cosmological distance measurements, is handled differently in the two cases. For whereas for Sandage and de Vaucouleurs, calibration is to be achieved through a methodological principle that one adopts at the stage of data analysis and interpretation, for HKP it involves calibrating the CCD cameras, i.e., the data collection devices themselves. My claim is that these differences can be explained by appealing to the concept of *material culture*, which played a major role in bringing the Hubble Wars to an end. For without taking the material context of the HKP measurement into the picture, in which a precision measurement is not obtained by methodological or 'philosophical' rigor, but by material interventions such as development of better software or introducing independent calibration methods on telescopes, their final result would not have been more than another discrepant number added to the list of Sandage's 50 and de Vaucouleurs's 100. It is the *material culture* that they operated with (and within), that made their ten per cent error claim – and the *epistemic iterations* that led to it – convincing enough for the scientific community to declare the Hubble Wars to be over.

Concluding Remarks

The measurement of the Hubble constant is still one of the active research areas in experimental cosmology and the debate over the value of the constant is far from being over.[54] Even though a discrepancy between the results of different teams still exists, scientists no longer talk about a 'war'. In this paper, I argued that this situation can be explained by observing that the *material culture* that Sandage and de Vaucouleurs worked with is no longer with us and, accordingly, its underlying epistemology no longer governs contemporary precision cosmology, even though the measurement and calibration issues are alive as ever. What ended the Hubble Wars was not only the existence of better measurements and data analysis techniques, but also the historico-epistemological context in which these activities took place.

10 MEASURING ANIMAL PERFORMANCE: A SOCIOLOGICAL ANALYSIS OF THE SOCIAL CONSTRUCTION OF MEASUREMENT

François Hochereau

Measurement is undoubtedly related to mathematics or statistics,[1] and more generally to scientific practice,[2] as the recent controversy over the reform of the international system of units has shown. But as Ian Hacking and Alain Desrozières have insisted, quantification and measurement did not emerge as a campaign of mathematicians or scientists; it derives in large part from the need of administrators for reliable information in order to be able to make plans to regulate societies and their activities. Even if the progress of science across the eighteenth and nineteenth centuries gives a larger role to measurement,[3] measurement has historically referred to practical activities in local markets, gambling houses, studios or shipyards.[4]

Witold Kula[5] rightly reminds us that units of measure have been supported and conditioned by social practices for a long time, referring to various amounts depending on the location, the time and the objects involved. With the advent of international system units, the benchmarks of local transactions moved from being representational measures to purely conventional ones. Measures were no longer based on units of counting and adding that everyone could understand (the foot for expressing differences between rows of potatoes, the step to assess distances and the cubic to evaluate the length of material). Instead, measurement became an issue of categorization, of means of reporting things with a normative dimension to define both what is measured and how it is measured through the establishment of a common system of units. Like standards,[6] measurement is part of a real infrastructure that invisibly shapes economic, political and moral practices.

Whether with regard to the social construction of measurements and measurement standards,[7] their calibration,[8] the intercomparison of the instruments,[9] the construction of valuation of measurements[10] or the normative power of measurements,[11] social science and historical studies have helped to shed light on the construction and the social impact of such measurement infrastructures. The goal of these social studies is to make these infrastructures visible, to subject them to analysis, to question their evolution, to explain their framing of social activities and to consider how these latter may sometimes resist to the measurement infrastructures' effects.[12]

To grasp the reality of measurement, we need to address the issue of its genesis and development in order to open the 'black boxes' of the social processes that underlie the institutionalization of measurement. To extend the debate on the construction of measurement to the focus on hard-to-quantify objects, such as emotions[13] or originality,[14] can be especially fruitful in helping to explain how objective knowledge is produced, either to define the object, means or purpose of measurement. Measurement involves different skills, artefacts, and representational and non-representational procedures that are dynamically interlinked.[15] Through this competition towards objectivity, the final measurement is then the outcome of negotiation and interaction between people and nature.[16]

We want also in this chapter to study the case of animal behaviour whose measurement as subjective one may be a good example of the tribulations of quantification to gain its objectivity through competing personal judgements and evaluations. The issue of measurement in animal selection is then to ask: what is the best cow? What requirements are debated? How to select animals? How to measure their performance? To what extent can we trust those measurements?

If one traces the recent history of animal measurement, we see the transformation described by Witold Kula, through the passage from representational or 'synthetic-qualitative' measures to purely conventional or 'abstract-quantitative' ones. In the past, the measure of the animal's performance was related to its social context. With the modernization of cattle breeding,[17] the performance became generic, indifferent to the specific qualities of each animal and to the relationship it has with the breeder.

The chapter proceeds as follows. After tracing the evolution of animal performance measurement through its transformation from a 'synthetic-qualitative' to an 'abstract-quantitative' form, we will focus on the emergence of a new kind of measurement in which it is the adaptation of cattle to their breeding environment that is to be taken into account. This new type of measurement is a good example of the transition from intensive farming, which optimizes all production-related parameters, to sustainable livestock farming, which tends to focus on the interactions between these parameters. Thus, farmers tend to focus less on milk or meat production than on animal robustness, disease resistance, autonomy and adaptability to various environments.[18] This evolution reverses measurement processes from reduction to extension in order to take various contexts into account. This recontextualization of measurement leads us to study the kinds of translation devices that are used to link local practices and official measurement codes.

The Evolution of Animal Performance Measurement

Up until the sixties in France, cattle were seen in three ways: as an economic rent because of milk or meat production, as a tool of labour and as a form of social and cultural heritage.[19] During this period, the value of an animal was assessed as a whole in terms of phenotypic characteristics (size, volume, dress...), aesthetic aspects (back line, head port...), and gait (stands, suppleness of walking). If the establishment of Herd Books at the end of the nineteenth century allowed collective agreements to define race standards, these standards focused mainly on characteristics such as color and morphology as a way of distinguishing each race from all the others.[20] Animal measurements allowed breeders to compare phenotypes according to specific morphological and aesthetic criteria within their own community of breeders. The difficulty of assessment lay in the fact that these measurements were partly conventional and partly representational. As exchanged in local markets and social heritage, the value of animals varied with persons, occasions and locations. For instance, there was no uniformity in value during animals sales in farm or local markets or during breeding shows. Animal assessments were strongly correlated to the reputation of breeders, the specificity of cattle breeding in territories and the social status of the parties engaged in relations of measurement.

From the 1970s onwards, a uniform national metrological regime was put into place for French Animal Selection. Faced with the challenge of breeding modernization in order to feed a growing number of people and make French agriculture competitive in the world market, animal selection had to be continuously improved through[21] the development of 'technosciences' and the diffusion of official measurement codes. The regime depended upon the development of quantitative genetics models to approximate animal performance and a large-scale data collection system tracking the performance of farm animals all around the country. The genesis of this regime is thus accompanied by the implementation of a far-reaching network through which measurements, calculation and assessments are computed to make more mechanized decision-making possible. The institutionalization of measurement values was supported by the technical and administrative ability of public expertise to manage data collection on cattle farms and diffuse standards of assessment all around the country, enrolling a wide range of actors such as extension consultants from the breeding associations, cooperatives of farmers, milk collectors and slaughterhouses.

This experimental network interfaced the knowledge production about animals and the control of animal choice as supported by a set of official measurement codes (called an index).

This second configuration embodies a displacement of the measurement activity from the animal as a whole to its intrinsic qualities as expressed by numerical codes. The emphasis on quantitative measurement moved cattle performance assessments from farms to technical stations gathering new kinds of expertise. The new metrological network also breaks the local animal representation into a synthetic one which incarnates an ideal type of animal performance. Thus, measurement is no longer defined through the comparison of animals but by counting and adding productivity parameters such as growth, height, weight, volume of meat, fat thickness, and so on. Measurement must satisfy a kind of economy of the metrological network; allowing a fracturing of qualitative things into quantitative parts that can be read and understood with minimal effort. Once the measurement categories are established by the metrological network, actors' assessments of animals begins to comply with them more and more, resulting in the adoption of these new standards via the power of a 'mechanical objectivity'.[22] As Theodore Porter[23] has explained, quantification is a 'technology of distance by diffusion of public knowledge', which is diffused everywhere on farms through the 'power of a discipline' that is more or less voluntary.

It was under the impetus of creating a common measurement for cattle performance that the modernization of the French livestock sector was achieved. This was done by translating each local measurement system into a universal, numerical language that could be accepted by everyone.

Over the past ten years, however, breeding goals have been challenged by both animal welfare[24] and environmental[25] concerns. This has led to a downplaying of the former economic view of animal productivity and the adoption of a mixed approach to animal performance assessment. Thus, a new way of evaluating animal performance, taking into account its adaptation to or robustness within its breeding context, needs to be found. In fact, this evolution reflects a return to old breeding practices when productivity was challenged by resistance to diseases and animal ability to grow by its own.[26]

To study this evolution, we have focused on the human–cattle relationship as a new kind of breeding criteria. With the radical changes in extensive breeding, cattle docility has become a major issue. Prior to the modernization of breeding (embodied in the second configuration of measurement), contact between breeders and their animals was frequent due to the close dependence of animals on humans for food, drink or grazing. Animals were handled daily by a variety of people on farms (parents, children, grandparents).

Today, the expansion of the animal population in individual farms parallels a decrease in the number of people who care for them, while the development of outdoor breeding during summer and the automation of cattle breeding during winter has led to a dramatic reduction in human–animal contact. The limited interactions that remain, moreover, are often negative (dehorning, prophylaxis, etc.). Thus, animals may often suffer stress when they encounter humans while humans take more and greater risks when handling animals.

In fact, the importance of docility issues arose with the setting up of the global testing network that supports animal indexation. Since breeders tend not to say much about the temperament of the animals they send to the public station, technicians may have to deal with unexpectedly aggressive animals. The animals suddenly find themselves in an unknown environment that more readily reveals their possible reactivity (which could be reduced by a strong relationship with their breeder).

In a certain sense, the scientific selection associated with the metrological network was questioned by a criterion that reintroduced the singular context of breeding into the generic selection system. The purpose of the metrological system and the quantitative statistics models was precisely to erase any farm effects in the index calculation. If we consider the docility issue, it affects the reputation of public selection by the diffusion of bad-tempered animals. Therefore, docility has always been a main characteristic of animal domestication, because it is a condition that must be satisfied in order to keep an animal in the herd. Moreover, the diffusion of some aggressive animals creates a risk for humans by increasing the accidents on cattle farms from animals that are too aggressive. In fact, the National Farming Health Service was a strong advocate of the introduction of docility criteria into the public selection system in the 1990s.

But how does one create a generic measurement of docility when the concept is embodied in traditional breeding practices based on the intimate relationship between farmers and their herd? If we adopt Kula's analysis, creating a measurement follows a process of abstraction in which certain traits are considered more significant than others in order to compare things more effectively. In fact, some breeders emphasize the impressive presence of their bulls in breeding shows, and so choose energetic animals that are full of character. But that kind of selection creates animals that are difficult for others to handle. This tendency is accentuated by the public selection system, which favours quantitative production traits to the detriment of more qualitative ones such as docility or animal behaviour. Making a more aggregated measurement requires collective agreement on the appropriate assessment of animal qualities and how they should be recorded. As explained by François Dagognet,[27] measurement is fundamentally relational; it is a universal language that allows everyone to share common views about something.

In our case, measurement reveals differences among the points of view of the various actors engaged in animal selection:

- The artificial insemination cooperative is concerned with their return on investment for the bulls they select for sperm diffusion on farms. To select a star breeder is extremely costly and time consuming. Because docility testing of progeny comes at the end of the selection process, striking an animal from the list may lead to a great loss of money.
- The traditional breeders focus on animals that have a noble bearing and a fine figure, which is often correlated with a strong temperament; on the other hand, bad temperament is often associated with poor domestication and relationship with the breeder. A single docility measurement could impact traditional breeders' reputations.
- The public breeding research station distributes sperm throughout the country and abroad. The expansion of animal populations in a variety of situations makes it necessary to select bulls whose offspring can be handled in any breeding context. This creates new challenge for the reputation of different cattle breeds and their economic development.
- The cattle farmers are concerned with the safe handling of animals and with minimizing working time with herds.
- The National Farming Health Service (FNHS) is concerned with the increase in accidents and their impact on workplace health and safety.

Faced with these diverse assessments of docility, the construction of a common measure will be based on a series of negotiations as to the definition of cattle docility and the encoding procedures by successive layout or formatting in order to make the docility measurement comparable. To enter into negotiations, the FNHS will ask an ethologist to propose an objective docility measurement supported by scientific knowledge. Ethologists from INRA (the French National Institute for Agricultural Research) were closely involved in the creation of that measurement, in part because it could strengthen their authority in animal breeding and within their own institute. At this time, they were marginal in comparison with other scientific fields: on one hand, animal behaviour was not central for agricultural professionals to increase herd productivity; on the other hand, ethological knowledge was considered to be less robust than that of other scientists. To gain legitimacy, an ethologist would first try to translate the reality of the breeding situation into the docility test in order to make the test relevant for future users. But the challenge was also to construct a scientific test, either standardized or reproducible, to ensure its generalizability across public breeding stations and cattle farms.

We can also describe the challenge of social acceptance for a new measurement as proceeding through four steps of legitimation: a test of reality, a test of objectivity, a test of relevancy and a test of usability.

(1) The first test, the 'test of reality', can only be conducted in association with field expertise to define an appropriate measure of the object. To do this for

docility, ethologists have involved technicians specialized in animal behaviour. The results of the exchange yielded a docility test consisting of three parts:

- The animal is first separated from its congeners (to eliminate animal social factors);
- It is left alone for thirty seconds, then for another thirty seconds a passive man is present (to prevent human–animal interactions);
- For two minutes, the man attempts to move the animal to the opposite corner of the park from its congeners (to prevent other animals influence) and then to maintain it in this corner for thirty consecutive seconds.

In this way, the docility test evaluates the animal's ability to curb its herd instinct and accept human contact. The evaluation of docility proceeds by adding other cattle behavioural measures such as locomotion activity and flight movements (reflecting the level of fear), and the time spent in the corner with the human handler (reflecting reactivity to the human constraint). In principle, this test appears to be reasonable objective, but it failed to interpret animal behavioural differences. In fact, measurement remains dependent up on the measurer and the singular context of the test. Even if the measurement protocol follows a scientific way, its results are too qualitative, not really relevant to compare numerous animals in a large population. Of course, all measurement is approximate: it can never be a complete reflection of the thing measured. But, as François Dagognet told us,[28] the strength of measurement lies in its incompleteness because it forces people to exchange and compare their results. Studying DNA typing, Linda Derksen argues precisely that these negotiations and interactions between people and measured things are precisely what build up the objectivity of measurement.[29]

(2) To overcome this 'struggle of objectivity', ethologists have called on INRA geneticists who are lead advocates of the genetic progress that supports the modernization of animal breeding. As such, they have an indisputable authority among ground-level expertise in animal breeding. Geneticists have managed with ethologists a large-scale investigation into the genetic inheritance of docility behaviour and the possibility of selecting animals based on this criterion.[30] The comparison of different cattle populations may now transcend the single cattle farm context to move to a generic situation, increasing the objectivity of measurement.

(3) Having acquired a method to grasp reality, and then having reinforced its perceived objectivity, the docility measurement must face another test before it may be used in breeding practices. This 'test of relevancy' queries the value of the docility measurement compared to other animal performance criteria. Although its designers claim that improving docility reduces the working time with animals and so the costs of breeding, this argument carries less weight than the market price of a bull at the end of the selection process. Thus, a stubborn

bull has been reclassified upward when compared to its superior zootechnical qualities. The docility measurement was then disqualified because of its irrelevance to the real context of livestock which favoured economic performances. The challenge was for the ethologists to work with cattle breeding professionals to incorporate docility assessment into their economic valuations. As a result, several tests have shown that docile cattle make good mothers for their calves (which leads to good growth and hence a better price); they are more resistant to diseases (which reduces drug costs); and they are easier to handle (which reduces production costs). Another experiment sought to study the interaction between docility behaviour and meat quality (in pigs, stress leads to sticky meat) but failed to find a significant correlation. All tests seek to verify the compatibility of the docility measurement with the other measurement systems used in animal breeding, such as those for calf growth, meat quality or disease prevalence. The measurement hybridizing docility, maternal behaviour and meat quality encourages reflection on the relationship between different measurement criteria and diverse local practices of animal performance evaluation. This mixed type of evaluation acts as a link between ground-level practices and the formal definition of measurement (which tends to reduce everything to a code). To a certain extent, the legitimacy of the docility test depends on its attachment to other measurements of animal performance.

(4) Even satisfying the different constraints of reality, objectivity and relevancy, docility measurement has to be easy to use in order to be well diffused and adopted. In fact, the test remains too complex to be used regularly either at public breeding stations or on cattle farms. Ethologists thus must develop a new, simpler test by which only movements are used to assess cattle docility in a normal situation (independent of the specificity of each farm and each measurer). The situation of animal performance control provided by external (and neutral) technicians on-farm has been chosen. The test consisted of counting the number of animal movements in the weighing cage as a sign of aggressive temper. Breeders were confused by this reduction of animal behaviour to a simple code, without distinguishing between calm or agitated, slow or quick movements and without considering the animal's behaviour prior to weighing. They suggested using instead a simple docility test already in use by Australian breeders as more relevant for them. That test consists of an approach to each animal by a person, who then assesses the animal's reaction on a scale of one (compliant) to five (aggressive). As a result of these controversies, all stakeholders decided to adopt a mixed measurement approach including a quantitative test (number of movements made during in weighing) in combination with a second test that was more qualitative, based on an individual assessment of cattle behaviour.

The problem of measurement, then, is to deal with the nature of the object to be measured and the capacity of measurements to provide adequate knowledge about that object. Measurements let us grasp the object through a universal coding, but the object can resist this reduction, prompting a new way of measurement. Such a dialectic is part of the dynamic of an inquiry.[31] We may also consider that measurements are not only a form of knowledge by acquaintance with the object, they are data constructed from that object for the purposes of further inquiry as to its better measurement.

Through the description of the social construction of the measurement of cattle docility, we have also explained successive 'investments in forms'[32] in order to match measurement to the nature of the object and its context of use. By maintaining agreements among actors as to the validity of the docility measure and by affirming its legitimacy within a multi-interested actors' network, these investments in forms help redefine and stabilize the measured object (from cattle docility to cattle performance) within time and space. The work of measurement must be studied not as a myth of total rationalization but as an always unfinished process. Measurement must be thought of as partial, unequal and temporary. It is important to explain 'le sens de la mesure',[33] taking into account its force and limitations. Measurement is not simply a matter of counting and adding fixed units. We must both collectively identify what we measure and define a way to measure it.

11 THE MEASURE OF DEMOCRACY: CODING IN POLITICAL SCIENCE

Sharon Crasnow

Introduction

Coding, a widespread practice in the social sciences, is a way of representing abstract or latent concepts in order to produce a dataset in which such concepts figure. While coding is often associated with the quantification of what otherwise might be thought of as qualitative data, the data need not be represented in a quantitative form. Dichotomous coding or coding on a qualitative scale is also used in political science. If we think of measurement as involving three elements – identification of the concept to be measurement, identification of a metric to which that concept will be matched and rules or procedures through which the matching is to be done – the question of whether such qualitative representation counts as measurement becomes a question of whether a metric must be numerical. Since this is disputed, there may be some question about whether coding in political science always qualifies as measurement. However, political methodologists standardly treat coding as measurement and I will do so here.[1]

Questions of measurement are intertwined with questions of standardization and accuracy. But standardization, a kind of uniformity, and accuracy both suggest a single end point of measurement and a single goal that measurement will serve. I argue that in the social sciences measurement of a concept like democracy using one standard may not be desirable and the sort of accuracy such a measure implies may not be appropriate. Consequently, an examination of measurement in the sphere of political science points to pluralism both in the aims of measurement and in the goals of research and knowledge production in which measurement figures.

Briefly, plural measures may be desirable since decisions about how best to code, or measure, social concepts may affect the conclusions that can be drawn from the resulting data. Minimally, transparency about such decisions is needed. We also may want to retain multiple datasets resulting from different coding (measurement) practices given that choices about coding are partially driven by multiple epistemic and pragmatic considerations. In addition, while measurement frequently is a step in testing hypotheses, the practices involved in developing a measure and the questions about the appropriate means to do so fulfil other research goals that sometimes are overlooked when we focus on testing as the main aim of research.

I use Hasok Chang's idea of epistemic iteration developed in his historical/philosophical work on measuring temperature as a resource to explore how measurement in political science involves pluralism both in relation to the goals of research and in relation to the plurality of activities that play a role in knowledge production.[2] In doing so I challenge the widely held presuppositions that we are seeking one answer to the question 'What is the best measure of democracy?' and that the goal of developing that measure is testing hypotheses.

Coding: Conceptualization

The social sciences (and political science in particular) use many concepts intended to capture political abstractions used in causal claims. Democracy is a good example. Concepts like democracy are not directly observable, and are often identified in the social science literature as 'latent concepts'. These are concepts for which observable indicators provide a means of measurement. Coding is the process of constructing datasets or indices based on these observable indicators. Coding involves both organizing and generating data in order to use them in analysis and inference. This is a necessary step for the statistical analysis of data, particularly multiple regression analysis, a popular tool in political science, but coding is also an aid to a more general organization of data.

Democracy provides a good opportunity for exploring measurement in political science. There are multiple indices of democracy generated using different coding practices and measurement levels and there are a variety of causal claims made within the field of democracy studies. For example, the empirical claim of a democratic peace – the claim that democracies do not go to war with each other – is supported through an examination of dyads of democracies.[3] Identification of these dyads requires the classification of states by regime type. Once classified, the dyad years can be examined and using different indices we might get different results. The democratic peace appears to be robust on a variety of measures, and so across different indices. Causal hypotheses intended to explain the democratic peace make use of these indices, as do other hypotheses in which democracy is thought to figure as a cause or effect.

Coding involves conceptualization – which in turn involves the selection of indicators – and rules for the operationalization of the concepts using those indicators. A number of political methodologists have written on these issues, and coding is typically discussed in social science methodology text books, such as John Gerring's *Social Science Methodology* where he lays out some of the key elements of measurement, claiming that it is 'virtually synonymous' with systematic data collection.[4] Gerardo Munck in his *Measuring Democracy* considers a variety of concerns that come up in coding democracy or creating indices of democracies.[5] He identifies conceptualization and measurement as key among these concerns. Thus I begin with a brief account of conceptualization and measurement using indices of democracy as my example.

With any latent concept – which is an abstract concept – the appropriate concrete attributes to associate with that concept need to be specifically identified. These attributes will either be observable themselves or linked to subcomponents which will be observable and serve as indicators. If too many attributes are specified, nations that ought to be considered democratic may be excluded (this would be a maximalist definition). On the other hand, specifying too few attributes risks including nations that should not be identified as democratic (minimalist definition). Consequently an index that was generated with either a maximalist or a minimalist conception of the latent term could result in a failure to adequately capture the causal relations under investigation, or, even if the index is appropriate to some range of phenomena, it may not be transportable.

For democracy, it is generally agreed that there are two main sets of attributes through which democracy should be identified: competitiveness (contested elections) and participation (inclusiveness of the electorate). Indices differ over the extent to which they focus on one of these attributes more than the other. For example, the ACLP (Alvarez, Cheibub, Limongi and Przeworski) index focuses on contestation (competitiveness) rather than participation.[6] According to Munck, ACLP argue that this is a legitimate preference since their concerns are really about the post-1945 era and so universal participation (universal suffrage) can be assumed and does not need to be treated as a significant attribute in these cases. The Polity IV index[7] also focuses on contestation rather than participation and given that the dataset reaches back to 1800, there is reason to think that comparison of emerging democracies and contemporary democracies could be problematic. Polity IV deals with this issue through a weighting of that attribute in the coding rules.[8]

This brief discussion of indices indicates that coders appropriately perceive an obligation to justify the choices that they make about attributes when coding. These examples also illustrate that the task of identifying appropriate attributes is informed by theory and background knowledge in an iterative fashion. '[C]onceptualization is both intimately linked with theory and an open, evolving activity that is ultimately assessed in terms of the fruitfulness of the theories it helps formulate ...'.[9] Questions of what factors are relevant and so what attributes are to be considered in the generation of the dataset are thus determined in part by theory, by arguments for attribute choice and, after the dataset has been employed, by its effectiveness (or ineffectiveness). While Munck emphasizes the role of theory in these choices, we should think of theory broadly – the background assumptions against which the theory is developed and accepted are also relevant, for example.

Another aspect of conceptualization that is informed by theory is the relationship between the broader concept and its attributes. The latent concepts are not themselves observable, but are associated with observations either of some attributes or by observations related to attributes in a specified way. Munck describes this relationship through a 'concept tree', where the more concrete parts of the tree are treated as the leaves of the tree. The observations that inform the coding are observations of these leaves. How well these observations capture the target concept thus depends on how well the concept is specified and how well ordered the concepts, attributes and components of attributes are. These features are well ordered when the relationships among them are clearly understood and described.

For example, if democracy is the target concept then one of the attributes identified is contestation (competitiveness). Two subcomponents of that attribute might be the right to form political parties and the freedom of the press. Each of these would be associated with particular 'on the ground' observations. Ultimately decisions about which subcomponents are to be treated as leaves of the tree and so indicators – what observations support that there is a free press or the degree to which it is free – need to be made. Each of these steps involves judgements about the subcomponents that have implications for the conclusions that the dataset can support.

Even if the concepts are seemingly well specified, questions of their consistency across contexts can arise. The term 'universal suffrage' did not mean the same thing in 1890 that it meant in 1973. Questions may also arise about whether the concept is used in the same way for cross-country comparative research. Decisions about how to treat such context-dependent indicators require careful attention given the potential for such differences to affect the extent to which the data support the conclusions.

For example, Pamela Paxton looks at what happens when universal suffrage is understood as universal male suffrage as it frequently is in the literature.[10] She argues that a number of standardly accepted 'truths' about democracy and democratization do not hold if universal suffrage is treated as universal male suffrage. It may also happen that while the concept and its attributes are agreed on – for example, it is agreed that universal suffrage is typically taken to mean the participation of all adult native-born citizens – when the actual coding takes place, background assumptions – such as, adult male suffrage is equal to universal suffrage – may inform the coding. This has ramifications for our understanding of democracy through these indicators in a variety of ways. Paxton identifies three: '(1) scoring dates of democratic transitions (2) descriptions of the emergence of democracy, and (3) research on understanding the causes of democratization'.[11] In addition, she notes that:

> when women are included in measures of democracy, the notion of waves of democracy (Huntington 1991) is no longer strongly supported. Neither does the West have a hold on early democratization. And, by the new definition, some countries' transition to full democracy may have come 30 to 70 years after the traditionally accepted date. The omission of women from our measures is therefore a problem with potentially far-reaching theoretical and practical consequences.[12]

Thus the specification of indicators affects typologies that are dependent on that data and hence may also affect reasoning dependent on these typologies. Such consequences are not always fully transparent, as Paxton points out, since even when researchers are aware that coding is not accurate in some sense – for example, they acknowledge that the understanding of 'universal suffrage' is not actually universal – they often may make a further assumption that the inaccuracy is not relevant.

This brief sketch of some issues for coding suggests one of its important elements – concepts – begins with theory (theories of democracy), but theory is itself based in observation. Conceptualization shapes the dataset but the dataset may be deemed inadequate after further observations that can, in turn, result in a revision of concepts. And it may also be the case that what we want knowledge for leads us to reconsider the attributes that are relevant to the concept. These two reasons for revising concepts are not the same. All representations require judgements about which attributes are relevant and those judgements have to do both with the accuracy of the representation and with its usefulness. Determining when we have it 'right' –when the indicator concept has both of these virtues (and perhaps others) – is an iterative process. The iterative aspect of conceptualization and the resultant coding is something that at least some political scientists have acknowledged.

Robert Adcock and David Collier offer an account of conceptualization that distinguishes what they call a 'systematic concept' from the background concept. The background concept is an unanalysed, 'ordinary use' concept and is often both vague and ambiguous.[13] Such concepts have been referred to in a variety of ways in the philosophical literature. Sometimes they are called 'cluster concepts'. Nancy Cartwright has called them *ballung* concepts, adopting a term used in Otto Neurath's work (*Ballung* is a German word that can be translated as 'congestion').[14] The idea has similarities as well to Wittgenstein's notion of family resemblance where there is no one set of necessary and sufficient conditions that determines the appropriate application of the concept term but rather overlapping properties and differing resemblances through which membership in the concept's reference class can be established. Other social scientists have suggested thinking of these concepts through fuzzy sets.[15]

Each of these approaches acknowledges that we begin with a concept that does not have a clear or 'scientific' meaning and that the concept needs to be made clearer in order to be useful in knowledge projects. Adcock and Collier describe this clarification as the production of a systematized concept. The systematization comes about through reasoning about the background concept and making decisions about which meanings are relevant for the research goals. The systematized concept is agreed upon by a community of scholars and commonly includes an explicit definition.[16] They emphasize the iterative nature of the process, quoting Abraham Kaplan's *The Concept of Inquiry*:

> Proper concepts are needed to formulate a good theory, but we need a good theory to arrive at the proper concepts ... The paradox is resolved by a process of approximation: the better our concepts, the better the theory we can formulate with them, and in turn, the better the concepts available for the next, improved theory.[17]

While Adcock's and Collier's discussion is useful, we will need to explore a little more carefully what 'approximation' means in the case of democracy. On the face of it, 'approximation' does not appear to have the same meaning as for physical properties that we measure – temperature, for example. In the case of temperature we are approximating to some physical property, however what are we approximating to in the case of democracy? Rather than approximating to some physical property (or properties) we are perhaps approximating to some aspect of a social concept *and* to a way of understanding that concept that will be useful for a particular exploration of the role democracy plays in a broader social/political context.

In Hasok Chang's discussion of approximation in the case of temperature he puzzles over the problem of circularity. Is it a problematic or benign circularity? Adcock and Collier's description of measuring is of an iterative, self-correcting cycle and, consequently, only problematic in that it is somewhat complex. But their analysis includes the idea that with each iteration we get a more accurate concept. But what does this mean when we are dealing with the latent concepts of the social sciences?

In what I have described at least one aspect of the systematization of the concept is that it must be suited to the particular uses of the research. Some conceptualizations are better able to serve those ends than others. This analysis also allows for the possibility that there is more than one end for which we want a concept like 'democracy' and these differing goals may require a plurality of concepts rather than a convergence to one. If this is so then measuring democracy is quite different from measuring temperature where a core agreed-upon use ultimately dominates and corresponds with an agreed-upon conception of the physical properties that are to be measured. For social science concepts alternative conceptualization of indicators suggests alternate measures. If different concepts serve different goals then we should expect an irreducible plurality of measures.

Sophia Efstathiou has an approach to such concepts that seems to fit here. In discussing race, she introduces the idea of 'founded' concepts.[18] Efstathiou understands founded concepts in a pragmatic way that seems suited to a plurality of measures, such as I have suggested for coding in political science. She specifies that the concept be '*founded* in that scientific domain when it is *transfigured* in appropriate ways to support the uses to which it is put there'.[19] When a physical property like temperature is what we seek to measure there is a convergence of measurement that is partially a result of agreement on its use. While this is not completely independent of the physical phenomenon that the concept seeks to capture, it is not entirely determined by it either. But if there are multiple uses of 'democracy' the diversity of uses would also support a plurality of modes of measurement. The notion that iteration would produce a 'better' version of the systematized concept may still be legitimate – but 'better' is an evaluation based on more than one virtue – not simply accuracy or truth – but other epistemic and pragmatic virtues may be relevant. The weighting of these virtues might also vary across contexts. Adcock and Collier's discussion of iteration does not capture this.

Coding: Testing and More

In the previous section I argued that datasets vary in ways that do not allow us to determine their value through focusing only on approximation or accuracy. Good coding or measurement is to be judged by examining both the pragmatic and epistemic virtues of datasets. Because different contexts may require that different virtues are maximized and different research projects often define or are defined by differing contexts, there may be good reason for the persistence of multiple datasets.

There is another pluralism that thinking about measuring social concepts such as democracy highlights. The construction of datasets is often seen as step to be taken in the process of testing theories. And indeed datasets do serve this purpose, although they may serve others as well. The following example illustrates both types of pluralism, both by suggesting that for a particular research project extant datasets are not adequate and by illustrating how developing datasets as part of a research project has other benefits in addition to providing data for testing.

Recent work by Stephan Haggard and Robert Kaufman on democratic transitions[20] starts with two indices of political regimes: that developed by Przeworski et al. (ACLP), recently extended and updated by José Antonio Cheibub,

Jennifer Ghandi and James Raymond Vreeland (CGV),[21] and Polity IV.[22] The two differ in a variety of ways. They cover different time periods: Polity IV begins in 1800 and CGV begins in 1950. Measurement levels also differ. CGV is a dichotomous coding and Polity IV is a continuous ranking of states. Polity IV scores are calculated through composite indicators, AUTOC (Institutionalized Autocracy) and DEMOC (Institutionalized Democracy) producing a twenty-one point scale (0 through -10 or +10) derived from the weighted sum of a number of indicators. The range is from +10 (strong democratic) to -10 (strongly autocratic). Above 6 is standardly counted as a democratic regime (though there is some variation).

The indices also vary conceptually. CGV is a minimalist measure of democracy, taking contestation to be the key indicator of democracy, whereas Polity IV is 'thicker' and looks at a wider variety of indicators, many of which are weighted after making judgements about the role of context.[23] Cheibub et al. argue that their minimalist concept allows measures to be more objective and so provides a cleaner way of thinking through the causal roles of democracy.[24] They argue that more complex measures, like those used for Polity IV, include subjective elements (for example, the weighting of factors) and make it difficult to sort out what causal mechanisms are operating. CGV is also a procedural account – it identifies democracy through a set of processes (like elections) rather than through institutions. Neither of these arguments appeals to the accuracy or empirical adequacy of the measure, but instead they make reference to other virtues of the dataset: pragmatic virtues such as that it is better for studying causal mechanisms; or theoretical virtues such as that it is more consistent with a particular theory of democracy.

Apart from their usefulness for categorization, indices provide data for causal reasoning where democracy is one of the variables. Examples include the impact of economic growth on regimes, the relationship between regimes and civil war, the impact of democracy on economic growth, and the relationship between inequality and regime change (democratic transitions) – the topic of Haggard and Kaufman's research.

Haggard and Kaufman consider causal mechanisms included in distributive conflict models of regime change looking specifically at democratic transitions and reversions in the so-called third wave of democratization (Samuel Huntington).[25] They reconsider a particular hypothesis about the causes of democratic transitions proposed by Carles Boix[26] as well as Daron Acemoglu and James Robinson.[27] These researchers use game-theoretic approaches to hypothesize causal mechanisms that link inequality to regime type.

Boix argues that high income inequality supports authoritarian regimes whereas high income equality and capital mobility are characteristic of democracy. He further conjectures that high inequality supports autocracy and transitions to democracy are more likely as economic equality increases. Acemoglu and Robinson argue similarly that autocracy and inequality have a U-shaped relationship with high and low levels of inequality sustaining autocracy and democratic transitions most likely to occur at levels that are more equal. The underlying features of the causal explanation are that it is distributive conflict that plays the major role in democratic transitions and distributive conflict is unlikely to occur at extremes of inequality/equality. A non-formal sketch of the causal mechanism is roughly as follows. At high levels of economic inequality elites will have too much to lose and so will prefer to repress when lower income groups attempt to mobilize and pressure for redistribution. At very low levels of inequality the motivation for redistribution will remain low and autocracies will not be challenged. But in the middle levels, the cost of repression may appear to be too high for elites and they will concede. Elites will initiate policy changes that not only have immediate, but also future, effects in order for changes to be perceived as credible. Thus they expand the franchise.

In order to explore this hypothesis as well as consider alternative hypotheses and more complex causal explanations, Haggard and Kaufman create an inequality and regime change dataset of country years during the period from 1980–2000. Next, using their dataset, they show that while a majority of democratic transitions conform to the hypothesis under consideration – they appear not to be the result of distributive conflict – a substantial number of democratic transitions (30 per cent) occur at high levels of inequality – cases that are coded as distributive conflict transitions. While their results are consistent with the original hypothesis (the average effect supports the hypothesis), the high number of non-distributive conflict transitions suggests that there are other causal mechanisms functioning in democratic transitions.

Haggard and Kaufman claim that the reduced form panels generated through the minimalist indices of democracy do not provide the information needed to adequately consider – and I would add, not simply test – the Boix or Acemoglu and Robinson hypotheses. CGV and Polity IV code democracy on too few key attributes and consequently fail to capture crucial elements (attributes) of such transitions. Haggard and Kaufman construct their index through coding any country year that was coded as a transition or reversion in either CGV or Polity IV as a distributive or non-distributive conflict transition (a dichotomous coding) using what they refer to as an 'extremely generous definition of "distributive conflict" transitions'.[28] They use process tracing to examine each of the cases and code the transition type based on observations made in each case. Their methodology is the qualitative methodology of case study research – i.e., methods such as the examination of contemporaneous news sources, historical documents and interviews. These observations are the indicators used to apply their coding rules. They next aggregate across the population of cases (not a sample but the entire set) to form their dataset. In this way they seek to determine '... not only whether inequality is associated with regime change but also whether its effects operate through the particular causal mechanisms postulated in distributive conflict theory'.[29]

Although they are testing for the hypothesized causal mechanisms, the case study research done in constructing the dataset also allows Haggard and Kaufman to identify relevant variables, refine their coding and generate new hypotheses. These are all activities crucial to knowledge production. Alternative causal pathways through which transitions come about, including distributive conflicts that do not follow the proposed distributive conflict models, also suggest the possibility of alternative classifications of conflict (typologies). Haggard and Kaufman consider the usefulness of other potential typologies for types of conflict – for example, ethnic or regional inequalities – and thereby suggest other attributes relevant for coding transitions. Given that the stability of a regime (and so the likelihood of reversion) may be a factor in coding for democracy the consideration of such attributes has the potential to affect the coding of democracy.

Epistemic Iteration – Epistemic Progress

The Haggard and Kaufman example is complex and I have concentrated on only a few of its features – those that most clearly affect the suitability of datasets for particular purposes and those that indicate the importance of measurement as an element of knowledge production. With these features in mind I now consider to what extent the types of iteration that appear in political science measurement can be elucidated using the notion of epistemic iteration that Chang uses in his historical account of temperature.

Haggard and Kaufman begin with extant datasets, but they have in mind a particular project that these datasets are not entirely suited to. However, the cases upon which the extant datasets are based serve as a starting point and give them a foundation. In the process of constructing their dataset questions of how and why the cases were originally coded as they were emerge. These can be understood as questions about the accuracy of the measurement of regimes (and so democracy, as one regime type). In fact, Haggard and Kaufman return to a few of these cases and recode them. The reasons they give for recoding have to do with the concept/measurement concern – that there are attributes that need to be considered that have not been included in the systematized concept. All of this conforms to the iterative process that Adcock and Collier describe in their account of measurement, except that Adcock and Collier see the process as aimed solely at accuracy.

But the iterative aspects of this process do not seem to be driven entirely by the desire for accuracy. Something like Efstathiou's 'founded concepts' may give us a more useful way to think about what is happening here in that it emphasizes the way that concepts constructed for one purpose may be altered to serve another. Efstathiou calls our attention to the pragmatic virtues that our concepts must meet. A particular measure – a particular coding – may not capture the features of the case that are relevant to the goal for which the dataset is desired. Its failure may be an epistemic one, but it also can just fail to be suitable as a tool to achieve the desired end. Specifically in this example, Haggard and Kaufman argue that a more fine-grained analysis is needed to understand why (and how) a causal mechanism functions in some cases but not others. There is a disanalogy with temperature, perhaps because there are multiple goals that we need our social science concepts to serve and they are more diverse than those that we require of our physical science concepts.

Chang characterizes epistemic iteration in the following way:

> Epistemic iteration is a process in which the successive stages of knowledge, each building on the preceding one are created in order to enhance the achievement of certain epistemic goals. In each step, the later stage is based on the earlier stage, but cannot be deduced from it in any straightforward sense. Each link is based on the principle of respect and the imperative of progress, and the whole chain exhibits innovative progress within a continuous tradition.[30]

He goes on to clarify how he thinks this works. We start by affirming some existing body of knowledge. This affirmation is in accordance with the 'principle of respect' which does not require that we wholeheartedly embrace that knowledge, but only that we take it as established and so a legitimate basis from which to begin. 'The initial affirmation of an existing system of knowledge may be made uncritically, but it can also be made while entertaining a reasonable suspicion that the affirmed system of knowledge is imperfect'.[31] The principle of respect is evoked to acknowledge the affirmation of the starting place. But we move beyond the original affirmation led by another aim, which is what Chang calls the 'imperative of progress'. Chang argues that to progress is not a matter of maximizing just one epistemic virtue. To progress is a matter of maximizing epistemic virtues more generally. He has in mind commonly agreed upon epistemic virtues – simplicity, empirical adequacy, fruitfulness, explanatory power, etc.[32] However, it is typically not possible to maximize all epistemic virtues at the same time for a variety of reasons. Some may be preferable to others for a given knowledge project – different virtues may be more important at different times, in different contexts and relative to different goals. Where we continue to have multiple goals –as it seems likely that we will in the social sciences – it seems quite plausible that we may continue to want multiple measures. In this at least some social science concepts will be unlike temperature.

In the Haggard and Kaufman example, the starting point is delineated through existing datasets and the theoretical understandings of democracy that inform those datasets and a hypothesis about democratic transitions. Conceptualization – part of the process of coding – depends on elements of theories of democracy, accumulated understanding of cases of democracy and other shared background knowledge from the field. Research starts with an affirmation of some aspect of knowledge – affirmation in the sense that Chang uses the term. The affirmation is provisional and this start of inquiry includes contested concepts, theories, and methods – we start with them because we need to start somewhere. The 'imperative of progress' is an imperative driven by the various epistemic (and pragmatic) virtues that shape research and guide the iterative process.

In Haggard and Kaufman the examination of cases indicates that the model is not empirically adequate (accurate) but there is a tension between empirical adequacy and explanatory power. Exactly how a transition occurred in *this* case may vary considerably from hypothesized causal mechanisms but the explanatory power of the causal models is not something that can be abandoned without good reason. Deciding which virtues to promote is not determined by a straightforward process of testing and correction either of theory or indices (coding) but a complex back and forth, where each stage is informed by the others, background assumptions are re-examined and the relative merits of the virtues are weighed in the context.

The measure of democracy remains an ongoing project and we do not yet know to what extent it may converge to one or more standards. If it does, that convergence will signal an alignment of both epistemic and pragmatic values. But it is also important to remember that while current datasets may serve as tools for testing, the decisions that go into their construction remain open to contestation. While this may seem to be a failure of the project of measuring democracy – all measures appear inadequate relative to some set of virtues – it could also be seen as a call for heightened awareness of the processes that go into the construction of datasets and the effects that these choices have.

Acknowledgements

I am grateful to Stephen P. Turner for his close reading and feedback on an earlier version of this work. Thanks also to Stephan Haggard and Robert R. Kaufman for explanation of and conversation about their research. All errors are mine alone.

12 MEASURING METABOLISM

Elizabeth Neswald

Introduction

In 1919, Arthur Harris, biometrician at the Carnegie Institute for Experimental Evolution, and Francis Gano Benedict, director of the Carnegie Nutrition Laboratory, published an essay on 'Biometric Standards for the Energy Requirements in Human Nutrition'. It begins with the general declaration: 'One of the primary requisites of all of the exact sciences is the establishment of standard bases of comparison'.[1] Formulated as a general statement on standardization and science, Harris and Benedict envisioned their statistical treatment of metabolism data as 'a prototype of that specialization in methods and coöperation in problems' which, they claimed, 'is the only means by which science can attain both the height of refinement of measurement and analysis and the breadth of comparison and interpretation which is essential to continued progress'.[2]

In 1919, when these lines were written, Benedict could look back on over half a century of attempts to standardize metabolism measurement methods, procedures, apparatus and subjects and at the contributions he and his laboratory had made over the past thirteen years to establish metabolism norms and promote standard instruments and methods. The need for standards and the difficulty of establishing them had frustrated nutrition researchers since the mid-nineteenth century. Human metabolism was difficult to measure. It required a high degree of technical expertise and skill and placed heavy demands on both experimental subject and experimenter, particularly in its early phase.

This essay sketches the process of measurement standardization in late nineteenth- and early twentieth-century metabolism research. Identifying steps in this process and pointing out the ways in which the methods and the objects of measurement and standardization revised and redefined one another, it proposes that the history of metabolism measurement can serve as a case study for standardization processes in an emerging scientific field. The essay begins with the phase of exploration and the beginning of measurement stabilization. Physiologists needed, first, to develop techniques and apparatus to accurately measure metabolism, and, in particular human metabolism, which required different apparatus and approaches than the measurement of metabolism in small animals or livestock. Once accuracy had been established, it became possible to consider ways in which metabolism measurement could be made simpler, more stable and thus more routine. This phase was shaped by questions of what methods, instruments, apparatus, procedures and techniques gave the most reliable results. In addition, since there were no standard apparatus, methods or procedures for metabolism measurement until well into the twentieth century, a means needed to be developed to make the results of different laboratories comparable. As physiologists became increasingly confident about their results, they began to debate what constituted an appropriate balance between measurement accuracy and measurement ease and whether a limited number of very precise measurements or a large number of sufficiently accurate measurements would provide more useful information for research and applications.

With increasing measurement stabilization and the accumulation of metabolic data began the search for predictors. This phase required the development of additional 'apparatus of commensurability'. By 'apparatus of commensurability' I mean the various instruments that were used to generate a basis for comparison, including material insturments and various kinds of statistical instruments, such as social surveys, means and averages, probabilities, indices and correlations. These means reached beyond the standardization of measurement procedures to the establishment of metabolic norms based on correlating physical factors. Through norms and a biometric correlation apparatus, researchers and clinicians could assess whether the metabolism of an individual fell within expected norms or showed abnormal or pathological characteristics. Metabolism measurement became routine. Finally, as questions of standardization and norms seemed largely resolved, metabolism researchers expanded their range of measurement subjects. In the process they found that their norms were not universal, as they had assumed, but were, instead, limited in scope and highly specific. After decades of refining their methods and norms, they were again faced with the question of what constituted normal metabolism and how it could be determined. This essay suggests that this trajectory from increasing complexity through exploratory diversification to experimental reduction, standardization and, finally, to increasing insecurity, as the area of application expanded, is not unique to the history of metabolism measurement, but that metabolism measurement can function as a case study for the process of measurement standardization in other fields.

Standardizing Instruments and Methods

Developing Measurement Instruments and Increasing Complexity

While some early metabolism research has been included in histories of nutrition and experimental physiology, most notably by Kenneth Carpenter and Frederick Holmes,[3] a history of metabolism measurement remains to be written. A central difficulty lies in defining historically what constitutes metabolism research. Some scholars take a broader perspective and include all study of total intake–outgo relationships, while others view it more narrowly as emerging from nineteenth-century physiological chemistry.[4] For the purposes of this essay, I will take the narrower view. From the late eighteenth century through to the 1840s, a number of researchers developed calorimeters to measure heat production in small animals and closed-circuit respiration apparatus to measure their carbon dioxide production and also attempted to devise methods for measuring the respiration of humans with face masks.[5] Despite some conjecture it was, however, unclear how or even whether the two phenomena were related.

In the 1840s, organic chemist Justus Liebig developed a chemical theory of nutrition that brought these two threads of experimental research together theoretically, albeit not experimentally. In Liebig's theory, carbohydrates and fats were the carbon-based substances that supplied the body with the fuel necessary to generate heat. They were combusted in the body in proportion to inhaled oxygen. Heat production was thus the product of respiration.[6]

Liebig's theory provided the basis for the research programme initiated by his pupil, Carl Voit. Voit began studying protein metabolism while an assistant to anatomist Theodor Bischoff.[7] In the early 1860s, he collaborated with hygiene professor Max Pettenkofer, who built a room-sized open-circuit respiration apparatus.[8] Whereas previous closed circuit apparatus had only allowed for short-duration experiments with small animals, this larger respiration apparatus enabled them to conduct experiments over several days on dogs and human subjects.

Pettenkofer and Voit began studying their first human subjects shortly after the apparatus had been completed.[9] The first respiration experiments were very elaborate. In the first years these experiments were intended to serve as a control for nitrogen balance experiments and thus often included protein metabolism analyses. In their studies on dogs, Voit and Bischoff had developed sampling and analysis techniques for the food fed to the animals and precise collection and analysis techniques for their urine and excrement. They used this to estimate the amount of carbon the animal consumed and excreted. Using human subjects added complexity. Voit often complained about the tediousness of meat preparation and sampling for the canine experiments, but preparing a chemically analysable mixed diet meal for a human was a complicated task that required careful raw material selection and cooking preparation so that the quantities of fat, carbohydrate and protein could be precisely determined.[10] Measuring human metabolism in the chamber was similarly more complex, due to the different needs of a human subject. Humans who were in the apparatus for extended periods wore clothing and needed bedding and something to do. These water and heat-absorbing objects could affect the results of the experiment, so their potential effects needed to be assessed. The experimental subject then spent at least twenty-four hours in the airtight sheet-iron chamber so that measurements covered diurnal and nocturnal metabolism and the period presumed necessary to digest the consumed food. Two steam engines drove the ventilation pump. Incoming and outgoing air were sampled and analysed, and outgoing air was saturated with moisture and passed through a gas meter. Various other devices and pumps were used for additional measurement controls and sampling.[11]

Although Voit and Pettenkofer assumed that the energy conservation law applied to living bodies, they did not have an energy-based nutrition theory. Their experiments, which set experimental standards in respiration gas analysis for the next decades, were based on the Liebigian theory of metabolism as a process of material exchange. They were, however, to use Frederick Holmes's terms, 'intake–output' experiments.[12] What went into the body and came out of it was measured or calculated as far as possible, but the internal processes that took place between intake and output were not the object of study.

Several years of experiments showed that the results from respiration gas analysis and nitrogen balance analysis largely agreed, and that for most purposes analysing respiration alone was sufficient to establish metabolism rate. Food, urine and excrement no longer needed to be analysed. While this simplified measurement in one regard, the apparatus itself became more complex, when Max Rubner, a student of Voit's, combined the two earlier instrumental approaches to metabolism measurement to build a respiration calorimeter for dogs in 1883. Rubner's apparatus consisted of a central holding chamber surrounded by an air chamber, which was contained in a heat-isolating water bath. Heat production was measured by changes in air temperature; carbon dioxide production was measured by an attached respiration apparatus.[13] While Rubner at first used an additional calorimeter to check the experimental one, he later integrated the control calorimeter into the body of the experimental calorimeter. He determined the heat capacity of all parts of the apparatus and sealed the openings for barometers, thermometers, screws, ventilation tubes, etc., with rubber to make the apparatus both airtight and heat insulated. The chamber was attached to various additional measuring instruments such as gas meters and self-recording volumeters, regulator systems for hot and cold water, mercury pumps for air sampling and chemical apparatus for air analysis. A water wheel and later an electric motor powered the gas meter. This apparatus and its auxiliary devices measured carbon dioxide production, the temperature in the calorimeter, control calorimeter and surrounding water bath, air temperature near the apparatus, humidity in the chamber and experiment room and air volume flowing through the chamber. By 1899, when American physiological chemist Wilbur Olin Atwater and his collaborator E. Rosa designed a respiration calorimeter suitable for humans, the description of the apparatus, its auxiliary equipment, measurement devices, checks, calibration procedure and analysis methods filled a small volume, with the tendency toward even more ornate apparatus continuing.[14] These increases in complexity were viewed as means to increase precision and accuracy, as more potentially significant variables were more precisely measured and controlled.

As has become clear, direct respiration calorimetry was a difficult undertaking. The apparatus were large and elaborate, expensive and difficult to build, and they required extensive auxiliary measurement and control instruments. They needed careful and time-consuming calibration and demanded a high degree of skill and patience to use successfully. They were highly technical apparatus that could be afforded by few laboratories and successfully built and used by very few, highly trained and skilled individuals. The direct calorimetric experiments conducted with these apparatus had several functions. Rubner and Atwater developed their apparatus in order to obtain extremely precise measurements of energy balance and prove energy conservation in metabolism. Rubner also studied the effect of different diets on heat production. With continuing experimentation, however, it became apparent that indirect calorimetric measurements, in which respiration alone was measured, were accurate enough to determine energy balance.[15] In this regard respiration calorimeters became both experimental apparatus and control apparatus. Once a sufficient number of experiments had shown that the results of indirect calorimetry agreed closely enough with those of direct calorimetry, the latter was no longer necessary and the complexity of the apparatus and number of measurement variables could be reduced. Measuring carbon dioxide production alone or in combination with oxygen consumption was sufficient for most experimental and clinical needs.

Reducing Complexity

This reduction in complexity made metabolism measurement much simpler, since respiration apparatus were less expensive than respiration calorimeters and easier to build and use. Although the confirmation that indirect calorimetry was sufficiently accurate was not considered established until the early decades of the twentieth century, due to the difficulty and expense of respiration calorimetry, several physiologists began working on various types of respiration apparatus from the late 1880s onward. Metabolism measurement apparatus became increasingly diverse.[16] The number of types increased enormously in the early twentieth century, so that eventually nearly every laboratory had its custom-built or modified device with its unique set of parts, materials, tube lengths, breathing apparatus and auxiliary apparatus complete with a diversity of operating principles and chemical analysis methods. Some apparatus used a chamber of various sizes – a room, an enclosed bed, a box covering the head. Others used breathing masks with covered or plugged noses. Apparatus were designed to use atmospheric air or oxygen tanks; some analysed carbon dioxide production during the measurement, while others collected it in a bag for later analysis.[17] Experimenters took different sized samples, measured for different lengths of time and let their subjects recline in different positions. They used random, paid experimental subjects, hospital patients and medical students. Some experimental subjects were trained in mask breathing before the experiment.

This versatility allowed physiologists to expand the situations in which they did respiration studies from the narrow laboratory setting to hospitals, real and model workplaces, high altitudes and different climate zones and to vary the kinds of experiments they did. Using various apparatus and set-ups, they measured metabolism and calorie consumption during typing and housework, agricultural and mental labour, playing the piano and riding a bicycle. With the increased ease of measurement, the number of studies conducted grew rapidly, and this increase in data was seen as compensating the loss in precision that had followed the transition from direct to indirect calorimetry and from laboratory apparatus to the pared-down portable versions.

Unsurprisingly, this variety of instruments, set-ups, techniques, subjects, conditions and operating principles could lead to large discrepancies, and results could often not be translated between laboratories. Such discrepancies could point to unanswered questions and open new research avenues or be the product of poorly conducted experiments, but they could also be simply an effect of the different apparatus and methods. Metabolism measurement knowledge was thus local, without standards that would allow the results of different laboratories to be made commensurable.[18]

Physiologists were aware of this problem of experimental and instrumental heterogeneity or, to use Elisabeth Crawford's expression, 'the hazards of cognitive fragmentation' that came from the lack of standard procedures and instruments.[19] They discussed it widely and introduced methods to combat it. When Benedict became director of the newly established Carnegie Nutrition Laboratory, one of his central early activities was to tour European physiology laboratories to view their equipment and methods.[20] The heterogeneity he found and the problems it caused are recurrent themes in his reports of these visits and were a central concern of the physiologists he visited. Some conducted experiments with different apparatus and used them to check one another by comparing the results and correcting for deviations. Others, such as Robert Tigerstedt and later Emil Abderhalden, wrote extensive summaries of the instruments in use or compiled multi-volume handbooks with articles on numerous instruments and methods.[21] With his interest in creating an international community of metabolism researchers, Benedict reacted to this heterogeneity in two ways.[22] A highly capable instrument designer, he developed a basic, flexible respiration apparatus, which he promoted both in the United States and on his European tours as a potential standard instrument.[23] Realizing that international standardization would take time and that local and national apparatus preferences could be persistent, in 1908 and 1911 he sent his assistant, Thorne M. Carpenter, to visit several European laboratories to inspect and train on their apparatus, which he then ordered for the Carnegie laboratory. Carpenter conducted a five-year comparative study of the most widely used European respiration apparatus and methods, testing their relative strengths and weaknesses and generating lengthy tables of the measurement results.[24] These tables presented typical deviations and effects and allowed researchers to compare experiments done with one apparatus or method with those done by another. Standard instruments, conversion tables, textbook codification and doubling experiments thus all belonged to the 'apparatus of commensurability'.

Developing Norms and Predictors

While measurement was becoming increasingly standardized, the subjects of measurement – the individual bodies – remained highly diverse. From the beginning, experimental nutrition physiology researchers were confronted with the variability of experimental subjects and their metabolism, and they struggled to find ways to categorize and assess individuals. Standardizing the subjects of measurement was thus the compliment to standardizing the techniques and apparatus of measurement, and both were a necessary prerequisite for compiling a basis of metabolism norms. Since metabolism measurement had a strong application orientation, these norms could be used as a clinical tool to predict the metabolism of a particular individual and evaluate deviations.

Due to this variability, in order to establish what rate of metabolism could be considered normal, researchers needed to construct a standard experimental subject. They did this in two ways: first, by introducing procedures that reduced the effect of contingent life circumstances and individual temperament as much as possible during measurements, and, second, by establishing correlations between physical factors and metabolism rate. Since the late 1890s, nutrition physiologists in a number of countries, primarily Nathan Zuntz and Rubner in Germany and Benedict and Graham Lusk in the United States, to a lesser degree Tigerstedt in Finland, had been pursuing research into the amount of energy the body needed for its internal processes, absent any external work and with only the unavoidable minimum of internal work. Zuntz introduced the term '*Grundumsatz*' for this minimum, which was translated by Lusk into 'basal metabolism'.[25] Such measurements were conducted on an experimental subject in a state of repose several hours after the last meal, so that neither external work nor the internal work of digestion affected the metabolism rate. Basal metabolism measurements revealed the basic running costs of unavoidable physiological processes such as breathing, maintaining internal temperature and circulating the blood. Beyond physiological interest in the minimum energy requirements of a healthy body, basal measurements were a reductionist move that made bodies more homogenous. Life conditions, environment, diet, physical condition and physical activity were so variable from one individual to the next that finding any kind of norm for metabolism required controlling for and, as far as possible, neutralizing what was social about the body and reducing it to a generic physical system.

Basal metabolism rates provided the basis for developing metabolism norms. Much systematic measurement took place at the Carnegie Nutrition Laboratory, the only laboratory in the early twentieth century that was dedicated to energy balance research and measurements. Its research programme aimed to establish normal basal metabolism from birth to death in both sexes. Within ten years, it had largely fulfilled this goal, and had also generated considerable data on energy consumption during a variety of activities, produced respiration apparatus for different situations, explored the effect of a number of variables thought to have a possible impact on measurement and compiled volumes of tables with conversion and standard values.[26] In addition, basal metabolism measurement methods themselves were becoming increasingly standardized. The twenty-four-hour period that Voit and Pettenkofer had considered the minimum period for accurate respiration analysis in their chamber had been reduced to a brief resting period in the laboratory followed by ten to twenty minutes of breathing into a respiration apparatus twelve hours after the last meal. Differences in procedures persisted along with differences in apparatus, but since the Carnegie laboratory had a clear leadership position, most basal metabolism data was generated by one laboratory, followed a single set of procedures and was done on similar apparatus.

While measurements made on resting, post-absorptive individuals reduced the effects of external activity and immediate diet, physical factors such as body size also affected energy requirements. The search for correlations between physical factors and metabolism rate, which would enable the compilation of prediction tables for energy needs, began with the shift from substance-based to energy-based nutrition theories in the 1880s. In 1879 physiologist Karl Meeh developed a formula for predicting body surface area based on the correlation between surface area and weight or volume.[27] Since bodies emitted heat through their external surface and heat production was an indicator of energy metabolism, surface area could function as a predictor of energy needs. Many physiologists assumed that living bodies emitted heat like any other bodies according to Newton's law of cooling, that is, in proportion to surface area and the difference between its temperature and that of the surrounding environment. A few years after Meeh established his constant, Rubner used it to compare the heat production of different animals, taking their absolute heat production and correlating it with body surface area in square metres. Since smaller animals had a larger surface area to weight ratio than larger ones, they had a larger cooling surface and higher relative energy needs. Rubner set this relationship as one of his biological laws.[28] Alternatively, other researchers calculated metabolism as heat per kilo body weight, in particular when doing group measurements, while many expressed their results in both correlations or translated between them. By the second decade of the twentieth century, both surface area and weight served as predictors of metabolism, while statistical analyses enabled the determination of the normal range of variability.[29] If the height or weight of a normal individual were known, his or her metabolism rate could be predicted with sufficient accuracy without it being measured, while the normalcy of a measured metabolism rate could be assessed by comparing it with the rate for individuals of similar height and weight. In short, metabolism rates could now be predicted and measurements interpreted using tables and slide rules.

Expanding the Area of Application

By the 1920s, metabolism norms had been established for both sexes from birth to old age, and clinical metabolism measurement was on its way to becoming routine. Research questions had also become routine: physiologists revised surface area measurement formulas, assessed the effect of minor variables on metabolism measurement and measured metabolism in special cases and diseases. A research programme seemed to have come to an end.

In 1924 Benedict collaborated with Grace McLeod of the Department of Nutrition at Teacher's College of Columbia University and Elizabeth Crofts of the Department of Physiology at Mount Holyoke College on a study of the 'Basal Metabolism of Some Orientals', which measured the metabolism rates of seven Chinese and two Japanese female students and compared them with the established norms. The results showed a number of anomalies, which led them to ask first, whether the standard metabolism correlation factors were applicable to these students, who had different physical proportions than their American peers. Second, they asked whether there might be a racial difference in metabolism, since these students showed uniformly lower metabolism rates than predicted.[30] This was not the first time that the metabolism of Asian individuals had shown lower rates than the established norms, but previous measurements had been made of Asians in Asia living under different cultural, climatic and dietary conditions than individuals in studies done in the United States or Europe. The students in the MacLeod study had been living in the United States for a time and consumed the standard student boarding house diet, however, which seemed to exclude environmental and dietary differences as factors.

The questions that this anomaly raised provided the impetus for testing the norms and predictors on a new set of measurement objects, or, perhaps more accurately, on a set of objects whose newness was defined by the measurement anomaly. Metabolism norms and predictors needed to be tested to see whether they were universal norms or, as Benedict and his collaborators proposed, merely 'American standards'. This question opened up a new research programme at the Carnegie laboratory to measure the metabolism of geographically and ethnically diverse and often arbitrarily defined populations.[31] Benedict's 'racial metabolism' research programme utilized the infrastructure of the Carnegie Institution of Washington. Researchers from other Carnegie departments trained at the Nutrition Laboratory on a simplified respiration apparatus developed for field purposes and measured, when convenient, the metabolism of local populations while travelling for other projects. Missionary teaching and medical networks in India and China provided additional researchers.[32] A number of researchers not directly affiliated with Benedict or the Carnegie Institution also contributed studies on the population groups in their regions, with some training on the apparatus at the Carnegie laboratory.[33]

These studies of the 1920s and '30s seemed to show that despite a large spectrum of variability, after correcting for outliers, Indian, Chinese and Japanese individuals had lower average metabolism rates than the 'American Standard'. A few groups such as the Maya and the Inuit had considerably higher rates, and some, such as the Jamaicans studied as part of Charles Davenport's infamous study of 'race crossing', showed no significant difference.[34] All results were, however, ambiguous, and measurement situations did not conform to procedural standards. The field apparatus proved not as robust as expected, experimental subjects were often frightened, distracted, malnourished or uncooperative, a second control measurement was rarely possible and measurements often took place in busy public spaces. In addition, the effects of climate, nutritional state and past diet could not be assessed with any degree of accuracy, and some studies conducted on foreign nationals who long lived in these regions found no significant difference between the immigrants and the local population. Nonetheless, most metabolism researchers assumed that racial differences in metabolism existed and that these non-Western populations deviated each in their own way from the established norms.

Criticism came from unaffiliated local researchers. A Japanese study from 1926 found that once the surface area formula, the main predictor, was corrected to better conform to the Japanese body type, differences between Japanese averages and the 'American standard' were smaller than differences between Japanese labourers and intellectuals. The deviation lay in the predictor formula, which was body-type specific, while diet had a substantial metabolic effect.[35] More pointedly, perhaps, the 'American standard' was shown to be itself problematic even for the supposed group of application. As a paper from the University of the Philippines criticized, it was just as unsound to group all European and American populations together under a so-called 'white race' as it was to speak of 'Orientals'. The American standard was itself derived from specific ethnic and sub-groups, mainly those of British and Northern or Western European heritage, under specific climatic conditions, i.e., in those temperate climates that had proven themselves most congenial to metabolism studies.[36] In short, expanding the area of application did not prove the universality of the standards. It destabilized them, showing them to be the product of measurements of a specific, limited group of objects conducted under specific local conditions.

Conclusion

As the preceding essay shows, metabolism measurement in the later nineteenth and early twentieth centuries went through several stages, beginning with the development of apparatus and techniques able only to measure a particular phenomenon – respiratory exchange. The early phase of metabolism measurement reveals interplay between elaboration and reduction. Apparatus and experimental procedures became simultaneously simpler and more complex, as the subject of measurement and the goals of measurement became more clearly defined and measurement variables were added or dropped. Measurement simplification had the immediate and seemingly paradoxical effect of increasing variability and hindering standardization, since more researchers had the skills to build and use these simpler apparatus, and the variety of apparatus and measurement procedures thus increased. By the end of the second decade of the twentieth century, however, a sufficient degree of standardization had been achieved to allow the establishment of metabolism norms. This standardization or, at least, measurement commensurability, was achieved in a variety of ways. One laboratory became dominant in this particular kind of measurement, which insured uniformity of apparatus, techniques and training in much research; one apparatus began gaining acceptance as a standard apparatus; researchers informally conducted comparative experiments and measurements with different apparatus; and these comparisons were systematically explored and developed and the results published.

Standardization of measurement required not only the standardization of apparatus and techniques, but also the standardization of measurement objects. Basal metabolism measurement fulfilled this function, as did the search for correlations between metabolism rate and other physical factors. Once norms and predictors had been established, metabolism measurement became a routine tool. More widespread use of the tool, however, revealed its limits and the unrecognized assumptions that had entered into its creation.

This sketch can only offer a preliminary overview of tendencies in the development of metabolism measurement, and more research is needed to explore the internal processes of development, such as how researchers negotiated their standards of precision and acceptable error and worked toward the standardization of procedures, as well as the role of social assumptions and applications in shaping measurement apparatus and technique development and measurement subject selection. These tendencies seem to indicate, however, that the history of metabolism measurement can function as a case study for the development of measurement procedures and their standardization and that it can be added to a growing body of literature on the history of metrology.

Acknowledgements

I thank the History of Science Department of Harvard University and the Max Planck Institute for the History of Science for hosting me during my research, Richard Lewontin for providing me a congenial workspace and the Canadian Institutes of Health Research for their generous support.

13 THE SOCIAL CONSTRUCTION OF UNITS OF MEASUREMENT: INSTITUTIONALIZATION, LEGITIMATION AND MAINTENANCE IN METROLOGY

Hector Vera

Introduction

The *International Vocabulary of Metrology* defines a unit of measurement as a 'real scalar quantity, defined and adopted by convention, with which any other quantity of the same kind can be compared to express the ratio of the two quantities as a number'.[1] From a sociological standpoint – rather than a strictly metrological one – what is crucial in this description is that units of measurement are 'defined and adopted by convention'. Someone may express this idea, loosely, by saying that units of measurement are 'socially constructed'. It is easy to say so meaning simply that units of measurement are not 'natural' entities that exist beyond the human realm, like asteroids or trilobites. But there is a more precise meaning which makes the idea of 'social construction of units of measurement' more helpful. In this narrower sense 'social construction' means that units of measurement are part of what common people believe to be real and what they take for granted while conducting their everyday life. The 'construction of reality' is part of the continuing human activity in the world and forms part of some of the essential dynamics that produce and reproduce social life.

This clarification is important because the expression 'the social construction of ...' has been greatly misused and overused, to the point that its initial meaning has been almost completely lost. In this chapter I will recover the primitive meaning of the concept meant to be part of a sociological theory of knowledge, according to the formulation made by Berger and Luckmann in the 1960s.[2] Based on that theoretical framework I will show in what ways units of measurement are socially constructed and how that construction shapes the form in which those units are actually formulated, used and modified.

Measurement and Social Construction

I will reconstruct here, in a concise manner, the theory of social construction of reality, closely following Berger and Luckmann's seminal work, paying particular attention to their sociological view on institutionalization, legitimation and knowledge.

The social order – Berger and Luckmann say – is a human product; it is created by men and women in the course of their ongoing activity. All human activity is subject to habitualization. Actions repeated frequently are cast into patterns that allow future actions to be performed in the same manner with economical effort. Habitualized actions become embedded as routines in the general stock of knowledge – knowledge that is taken for granted and is at hand to any member of a society. Whenever there is a reciprocal typification of habitualized actions by types of actors we are in the presence of *institutionalization*.

Institutions imply historicity and control. Institutions have a history of which they are products and they can only be understood considering the historical processes that produced them; simultaneously, institutions control human conduct by setting patterns of conduct. When a new institution is passed from one generation to the next they become historical institutions, and with historicity they acquire objectivity. Institutions are then crystallized and are experienced as existing beyond the individuals who embody them – they are perceived as possessing a reality of their own. At this point the institutional world hardens and gains massivity – this includes that institutions set and maintain certain meanings and definitions of reality that are imposed over the individuals participating in the institutional process.

The humanly created institutional world requires *legitimation*, ways by which it can be explained and justified. Legitimation produces new meanings to integrate the existing meanings of disparate institutions. It provides integration and plausibility to the institutional order – it makes that order look plausible to the participants of different institutional processes. There is a cognitive and normative element in legitimation. The former explains the institutional order and gives it cognitive validity; the latter gives it normative dignity to its practical imperatives. In other words, legitimation tells individuals why things are what they are, and why certain actions should be carried out and others should not.

Institutions are legitimated by living individuals who have precise social locations and interests. When legitimations become problematic some procedures of maintenance are necessary. Since all social phenomena are historically constructed through human activity – they are always 'in progress' – no social order is entirely taken for granted and all social orders are incipiently problematic.[3] Alternative definitions of reality appear constantly, and it is the work of custodians of the 'official' definitions of reality to put in action repressive methods to incorporate, quiet or annihilate the divergent definitions.

Social and conceptual machineries of maintenance work to sustain the institutional order and its forms of legitimation. The success of these machineries is related to the power of those who operate them. What definitions of reality will prevail depends more on the power of its carriers than on its theoretical relevancy or coherence. Power in a society thus includes the ability to set socialization processes that are crucial to *produce* reality; and when push comes to shove, definitions of reality could sometimes be enforced by punitive organizations, like police.[4] When these instances of reproduction and repression are highly successful, experts are able to hold a monopoly over the definitions of reality.

What can we learn from these theoretical principles to advance a sociological understanding of measurement? First of all, units of measurement are, obviously, socially constructed, but they are no constructed in the sense of being finished or having acquired a definite form, they change regularly. Like the social construction of any other 'reality', units of measurement are created and recreated by processes of institutionalization, legitimation and maintenance. The institutions of measurement are constituted by a long series of social actions that are habitualized and form part of a social stock of knowledge, which brings steadiness to the institution and reduces its malleability. When a conventional way of measurement is passed from one generation to the next it acquires objectivity and is subjectively experienced as possessing a reality of its own. The institutions of measurement need to be legitimated: someone has to explain why a particular way of measuring is pertinent and why measurement should be done in that way and that way only.

Historically, the legitimation of measurement institution is carried out by experts who are backed by powerful groups with clearly defined political and economic interests. These experts and groups try to inhibit the appearance or spread of any alternative way of measurement. If they accomplish that, they can hold a monopoly on the units, concepts, instruments, methods and systems of measurement used in a society.

There is, additionally, another aspect in Berger and Luckmann's theory that is crucial for the sociological study of measurement: the social accumulation and distribution of knowledge. When individuals of a given society share a common biography, a portion of their intersubjective experiences become sedimented[5] and are incorporated in a common stock of knowledge. This sedimentation is objectivated in a sign system (like language) and with it experiences become impressionable, transmittable and available to present and future members of a linguistic community, forming thus the basis of the collective stock of knowledge.

This stock of knowledge is structured and divided in what is generally relevant and what is relevant only to particular groups – there is thus a social distribution of knowledge. We find a body of knowledge that is more or less evenly distributed among all members of a society and sectors of the stock of knowledge that are only pertinent to certain areas of activity, which are administered by specialists.

The distinction between the general stock of knowledge and specialized knowledge is quite important for sociological analysis. It is not the same to study the knowledge of a specific group (like experts) than studying the knowledge that is shared by virtually all members of a given society. In the case of metrology, it is necessary to see mutual influences and restrictions between the specialist's knowledge and the general stock of knowledge. In the cases that will be considered here it is possible to see that sometimes specialists can revolutionize and reshape age-old elements of the general stock of knowledge (like with the introduction and imposition of the basic units of the metric system). But specialists also have to deal with severe limitations to their theoretical creativity and technical capability when they try to transform the units of measurement that people use in everyday life (e.g., the most extensively used units of time measurement: second, day, week, year). The broader social use of units of measurement has implications that cannot be avoided by the scientific organizations that define and regulate those units – as rule of thumb one can say that the wider the use of a unit of measurement by non-experts, the harder it gets to modify it.

When a scientific organization decides to make an amendment or redefinition in the units of measurement in their field, it usually provokes long and impassioned discussions among its members. These debates frequently unveil the values, conventions, traditions, interests and convictions of the participants. How these debates are resolved could determine for a while the direction of futures disputes in those communities of experts. These processes are of great importance for social studies of science; however, if we expand the viewpoint to see beyond the walls of laboratories and universities, we will find out that for the most part laypeople will not even have heard about what happens with the units of measurement defined and revised by scientists. This is so because experts are forced to work within very narrow limits to keep those units of measurement widely use in a larger society as an *invisible* infrastructure. It is almost impossible to achieve any truly radical change in a globally used measurement system (like introducing a thirteen-month year) because metrological conventions (and the institutions and behind it) are a tough nut to crack. When we consider the implications of the social life of measures, by observing the metrological practices and understandings of the population at large, we enter into the domains of the sociology of knowledge in its broader sense.

The tense coexistence of a 'folk' metrology among the masses and a scientific system of weights and measures among the elite of trained experts (whose bodies of knowledge are removed from the common knowledge of the society at large), both interacting in the institutional order of measurement, has been a constant since the invention of the metric system. These two analytical levels require distinct lenses to be grasped. Scientific metrology and the technical reformations of the unit of measurement belong to the field of the history of ideas; on the other hand, common-sense knowledge – not the sphere of scientific ideas and theoretical thought – is what is relevant for the sociology of knowledge, which pays greater attention to 'everything that passes for "knowledge" in society'.[6]

In the coming sections of the chapter I will illustrate these ideas with four cases of the social construction of four units of measurement: the metre, the kilogram, the second and the carat. Complementarily, the chapter considers how the interaction between experts and laypeople imposes severe limitations on any plan to transform units of measurement radically, as well as specific ways to legitimize and maintain the existence of present units of measurement.

The Decimal Metric System: Renovation by Revolution

The decimal metric system was invented in the 1790s by the French revolutionaries. They aimed to create a measurement system based in nature and that would be universally used. The base units of the system were the metre, litre and kilogram, which were meant to be 'natural standards', linked to unchanged natural phenomena. The metre was first defined as one ten-millionth of the distance from the equator to the North Pole; the litre was the volume of a cubic decimetre; and the kilogram was the mass of one litre of water at its maximum density. The system had a set of prefixes that function as decimal multipliers.

Regarding the aim to create a system 'For all *time* and for all *people*', things *have gone almost as they envisioned*. The worldwide diffusion of the metric system has made it the first global system of measurement. Nowadays roughly ninety-five percent of the world's population lives in a country where the metric system is the exclusive system of measurement – the latest adoption was by Myanmar in 2013, leaving only six other countries in the world not committed to the exclusive use of the metric system: Liberia, Marshall Islands, Micronesia, Palau, Samoa and the United States.[7]

This was an impressive achievement, especially considering that the metric reform was quite radical. Reforming a system of measurement involves changing one or all the nuclear elements of the system to be modified: (1) the magnitude, size or amount of the units of measurement; (2) the names of the units; (3) the system of grouping and division (e.g., base-10 or base-12). All combinations can take place in changing these elements. It is common that a unit varies in its magnitude but keeps its name and subdivisions. Also common is that a unit changes its magnitude and multiples but keep its name. And there have been a few particularly extreme reforms that change these three elements at once, like the metric system did. Its inventors decided to create units with new magnitudes, provided a novel nomenclature for those units, and obliterated duodecimal and sexagesimal divisions in favour of a decimal system.

But despite its global triumph, the metric system has not remained unchanged. In 1960 it was renamed as the International System of Units (SI) and it has now seven base units: metre, kilogram, second, ampere, kelvin, candela and mole. The litre, as a based unit, was lost. The metre and the kilogram, on the other hand, remain at the centre of international metrology, but they have gone through several adjustments. In 1799 the metre was first embodied in a platinum bar known as the *mètre des Archives*, which was replaced as official standard in 1889 by the International Prototype Meter. In 1960 the metre was redefined in terms of the light from an optical transition of krypton-86; and its most recent redefinition took place in 1983, this time in terms of the speed of light.

Regarding the mass standard, in 1799 a platinum prototype was fabricated to embody that standard and it became known as the *kilogramme des Archives*. In 1889 it was replaced by a cylinder of platinum-iridium, the International Proto-type Kilogram (IPK). The IPK still embodies the base unit of mass of the SI. The kilogram has been the least altered unit among the original units of the metric system. It remains the only SI unit defined by an artefact; but apparently this is about to change; in recent years the discussion about the redefinition of the kilogram has produced heated debates in the international metrological community[8] – more on this later.

Why Did Antecedents and Contemporary Competitors of the Metric System Fail?

Two questions have to be asked to understand the present institutional order in world metrology. First, how was the metric system able to displace hundreds of traditional measurement systems around the world? Second, why was the metric system, and not any other system, that one that became a global metrological language?

One of the great stories in the history of the metric system was the marginalization and obliteration of thousands of units of measurement around the world when the metre, litre and kilogram were adopted by country after country during the nineteenth and twentieth centuries. A lesser-known aspect of the process of global diffusion of the metric system is its confrontation against a number of measurement system created *ex professo* to challenge it. The overwhelming triumph of the metric system in scientific circles over those challengers helped its ultimate global success greatly, as experts in different countries were able to present plans for metrological reform centred on the metric system (thus avoiding internal disputes to define what system should be chosen to supplant customary measures). Since the second half of the nineteenth century the lack of national metrological uniformity and the absence of international coordination in weights and measures were afflictions that found everywhere the very same prescription: metrication.

The inadequate understanding of this phenomenon and the lack of research on who challenged the metric system with newly invented measurement systems has created the impression that those plans were not a factor in the history of metrication.[9] Some commentators have incorrectly argued that 'left without competition as the only scientifically based, decimalized measurement, the French metre has taken its present place as the simplest, most accurate means of measuring everything between the dimensions of a quark and a black hole'.[10]

This is a misconception. The metric system faced plenty of competition. Initially, at the end on the eighteenth century there were several plans for metrological reform that came about shortly before or at the same time as what the French savants were doing in the early 1790s when the metric was created. Just to mention a few of them, there were plans formulated by Cesare Beccar in Italy, by James Watt in England and by Thomas Jefferson in the United States.[11]

Why, among these different plans, the metric system did grow into a global language and the others became mere footnotes in historical metrology? The metric system had multiple resemblances to the others – and it was itself quite similar to a plan sketched by Gabriel Mouton in the seventeenth century.[12] So it is hard to believe that the metric victory was due to its purely technical virtues. What the metric system had was the fortune of being part of a *social* revolution.

Revolutions – and other forms of rapid social and political change – have been the midwife of metrication. This can be illustrated, for example, with the introduction of the metric system in Russia. Near of the end of czarist times Dmitri Mendeleev was director of the Russian Bureau of Weights and Measures and he pushed for the metrication of the country. But not even his influence as one of the greatest chemists of the century was enough to get that reform made. It was Lenin and the October Revolution that made that plan a reality in 1922.[13] Radical metrological reforms follow radical social change.

One of the central elements of the social life of measures in Europe and the Americas until the beginning of the nineteenth century was *multiple metrological sovereignty* – i.e., the lack of a single unified political hierarchy to regulate weights and measures and the presence of competing claims by opposing parties over metrological authority. There usually was a proclaimed sovereign metrological authority, but in practice that authority was ignored or challenged. This situation ended in France on 4 August 4 1789 with the abolition of feudal privileges that deprived lords of the right to have the final say in metrological matters in their own estates. The revolution unified all the French under a sole authority that set a single system and settled all disputes – of course it is not a coincidence that the birth of the metric system and the establishment of this sovereign metrological power occurred simultaneously.

In the nineteenth and twentieth centuries, when other nation-states looked for the proper instruments to secure this monopoly, the metric became a ready-made system, legitimized by its scientific aura, and with a proven record of success.[14] Once the metric system was established and started its global dissemination many other proposals were advanced to challenge it. These projects included new units of measurement and some were based on non-decimal systems of grouping and division (they proposed instead base-8, base-12 and base-16 systems). Just to mention a few of these published in the second half of the nineteenth century in Europe and the United States: J. W. Nystrom, *Project of a New System of Arithmetic, Weight, Measure and Coins: Proposed to Be Called the Tonal System, with Sixteen to the Base* (1862); V. Pujals de la Bastida, *Sistema métrico perfecto ó docial* (1862); A. B. Taylor, *Octonary Numeration, and its Application to a System of Weights and Measures* (1887); E. Noel, *The Science of Metrology or Natural Weights and Measures: A Challenge to the Metric System* (1889).

Even Herbert Spencer sketched out in 1896 the outline of a new system of measurement which he proposed to be used worldwide in the future instead of the metric system. Spencer claimed that duodecimal systems were greatly superior to the decimal metric one; the great divisibility of twelve made them more practical. He feared that if the metric system was to receive universal acceptance, the obstacles to change in the future would be insuperable. He wanted a profound reform that included changing the whole numbers system making it duodecimal. 'It is perfectly possible', he said, to have all the 'facilities which a method of notation like that of decimals gives, along with all the facilities which duodecimal division gives. It needs only to introduce two additional digits for 10 and 11 to unite the advantages of both systems'.[15] This was, of course, a farfetched plan that received practically no attention (even if the part of his criticism of the metric system was frequently repeated).[16] The other alternatives to the metric system were also mostly ignored.

One of the few thinkers who gave some consideration to one of these publications was the philosopher Charles S. Peirce, who at the time worked as metrologist at the US Coast and Geodetic Survey.[17] Peirce reviewed Noel's *The Science of Metrology*, and he showed great skepticism towards his plan – 'Mr. Noel's system is nearly as complicated and hard to learn as our present [English system], with which it would be fearfully confused, owning to its retaining the old names of measures while altering their ratios'.[18] And making a diagnostic of the present and future metrological situation in the United States, Peirce said:

The whole country having been measured and parcelled in quarter sections, acres, and house-lots, it would be most inconvenient to change the numerical measures of the pieces. Then we have to consider the immense treasures of machinery with which the country is filled, every piece of which is liable to break or wear out, and must be replaced by another of the same gauge almost to a thousandth of an inch. Every measure in all this apparatus, every diameter of a roll or wheel, every bearing, every screw-thread, is some multiple or aliquot part of an English inch, and this must hold that inch with us, at least until the Socialists, in the course of another century or two, shall, perhaps, have given us a strong-handed government. We can thus make a reasonable prognosis of our metrological destinies. The metric system must make considerable advances, but it cannot entirely supplant the old units. These things being so, to 'challenge' the metric system is like challenging the rising tide.[19]

This was an accurate explanation of why challenging the metric system was a naïve idea. The metric system succeeded in replacing hundreds of customary methods and units of measurement around the world that had existed for centuries. Provably equally surprising is the fact that no other modern and scientific measurement system designed in the nineteenth century was able to challenge it seriously. The metric system achieved then a double victory. We should explain this victory in sociological terms – and Peirce had a revealing intuition hinting at the need for a strong government to have metrication in a country.

States and the Social Organization for Maintenance of Metrication

During the nineteenth and twentieth centuries everything that is crucial to establish and preserve a functional system of measurement (e.g., the authority to define units of measurement, store physical standards, prescribe proper methods of measurement), was expropriated from the hands of various, uncoordinated authorities – cities, corporations, guilds, town markets – by central state authorities. In the same way that states dispossess their domestic competitors of the instruments of physical violence and the right to use them, so they warrant their monopoly on the legitimate means of measurement by dispossessing social groups of their measuring rights and authority. Briefly, this process can be described as the transition from *multiple metrological sovereignty* to a single sovereignty that holds the *monopoly of the legitimate means of measurement*. Usually, multiplicity of systems of measurements in a single territory reflects a multiplicity of sovereignties. That is why it is not an accident that the adoption of the metric system is, most of the time, preceded by a massive social change – a change frequently characterized by the destruction of old institutions and the start of a new central sovereignty, like revolutions, national unifications, colonization, and so on.

The rights that are included in the monopoly over the legitimate means of measurement include, among others, the right to: determine the legal units of measurement – and the subsequent banning of all other units – making those units the only acceptable in civil contracts and commercial exchanges; retain the physical standards in which the official units of measurement are embodied; implement a traceability chain to secure the faithfulness of all standards and instruments; resolve how frequently standards should be verified and who can verify them; determine *how* objects should be measured (this includes procedures of verification and the proper way to weight commodities in the market place); designate inspectors to validate proper employment of weights and measures; set fines and penalties for metrological wrongdoings; specify when and how the official measurement system should be taught in schools.

Establishing a monopoly on the legitimate means of measurement serves many purposes for a state; and, conversely, this monopoly has greatly helped the cause of the metric system. In other words, *modern nation-states need homogeneous systems of measurement, like the metric system*, and *metrication needs the states*. States need metrication because the establishment of the metric system gives them leverage to fulfil some of their essential functions: enhancing the extraction of revenue; making the population and economic resources 'legible'; monopolizing symbolic capital; undermining the influence of local authorities; consolidating internal markets; and sharing scientific and commercial standards with other countries. Metrication needs the state because only modern states have shown to be effective in helping, compelling and, if necessary, forcing populations at large to employ metric units; science, commerce and industrialization alone have not been able to do this. A successful metrication process, in other words, requires *compulsion,* and it can only be achieved on a large scale when two actions are combined: (1) policing the employment of metric units; and (2) providing populations with the intellectual and materials means to learn the metric language. And states have been the only body of institutions able to do all that simultaneously and effectively.

Scientists and the Conceptual Machinery for the Maintenance of Metrication

Once the metric system became an official system of measurement – i.e., an official definition of reality – it needed a conceptual machinery to displace or liquidate deviant definitions, which were neutralized by assigning to them an inferior ontological status and a 'not-to-be-taken-seriously' cognitive status.[20] Furthermore, in every social order there is a need for social apparatus to transmit institutional meanings from the people who know them to the non-knowers – in our case people who know how to use the metric system and people who do not. This transmitting personnel includes 'theoreticians' (who integrate into a coherent from the different institutional meanings) and 'pedagogues' (who help in the transmission of meanings to new generations in an easily understandable way).[21] In these various tasks the role of experts was crucial.

Experts were very efficient in presenting the metric system as the only 'rational' system that could be adopted in commerce and public administration; they gave to the metric system a scientific aura. See for instance the terminology used in the opening paragraph of the decree that made the metric system compulsory in El Salvador in 1885: 'the old Spanish system of weights and measures does not have a rational basis and it is then prone to inexactitudes and opens the door to continuous frauds. ... The French metric system, adopted already by the majority of the civilized nations and it is based on units that are found nature, has the characteristics of exactness and simplicity'.[22]

This definition of the metric system as *the* rational answer to metrological problems was self-reinforcing – as definitions of reality 'have self-fulfilling potency'.[23] This was particularly clear with the internationalization of the metric system. One of the most conspicuous aspects in the campaigns in favour of the metric system was the frequently repeated prediction that the global expansion of the metric system was 'inevitable'.[24] Governments, industrialists, merchants and professional associations were convinced that the metric system was going to became an international and exclusive language of measurement in the foreseeable future and its implementation was not a matter of 'if' but 'when' it was going to happen. The language of inevitability became ubiquitous in metric debates. A couple of examples of this would suffice: 'Why should we use the metric system'?, asked the physician E. Wigglesworth in 1878, answering '*Philosophically; as accepting the inevitable*, for the metric system is sure to come';[25] and in 1902 an editorial of the *New York Tribune* read 'Conformity to one common system of weights and measures is not only extremely desirable, but certain to come. The only doubtful question is how long it will take'.[26] These rather audacious predictions were usually defined as real – and they were real in their consequences. The assumed inevitability of the internationalization of metrication became a powerful engine for the advancement of metrication itself.

If the state apparatus was at the core of the social organization to push forward the adoption and utilization of the metric system, the conceptual machinery to maintain and legitimize the metric system was a task for specialists. The role the intelligentsia (scientists, engineers, educators, university professors and men of knowledge in general) was thus crucial in the metrication process, as they fulfilled three vital functions related to the social distribution of metrological knowledge: (1) through their communication networks they propagated information about the metric system and its advantages; (2) they provided states with the necessary expertise to articulate and materialize the adoption, management and teaching of the metric system; and (3) they articulated the legitimation of the metric system by entangling it with values such as 'science', 'civilization' and 'universality'.

Refurbishing an Old Unit: The Carat

The radical change of units of measurement attained by the metric system should not be considered the norm; it was rather an unprovable accomplishment. Usually units of measurement have great resiliency and it is difficult to change them drastically. This does not mean that units remain unaltered either; they usually vary gradually, adding minor adjustments through extended periods of time. This is how units of measurement that were invented in antiquity, like the carat and the second, are still well and alive.

The story of the carat – the unit of mass used for gemstones – illustrates the process of redefining and fruitfully standardizing a unit of measurement within a relatively narrow professional community and it is an exemplary case of how old units of measurement can be modernized without making a radical cut with the past. The carat is an ancient unit; its very name reveals much of its long history. The word *carat* is derived from the Sanskrit *quirrat,* which was a reference to the elongated seed pod of the carob, used in antiquity to weigh precious metals and stones.[27] The *Oxford English Dictionary* indicates that it was incorporated into English in the Middle Ages from Italian (*carato*), coming from the Arabic (*qirat*) and the Greek (*keration*).

As its etymology suggests, the carat has been widely used across times and cultures, but it is not an unambiguous unit of measurement. First, because the carat can be either a unit of mass for gemstones or unit of the relative purity for gold alloys (in English the spelling alternative *karat* is used to refer to this second meaning). Second, prior to the twentieth century, the carat as a unit of mass had quite different magnitudes – usually weighed in Troy grains – in various trade centres around the world: 213.5 mg in Turin; 205.8 mg in Portugal; 205.1 mg in Russia; 196.5 mg in Florence; 192.2 mg in Brazil.[28]

The activities related to precious metals and stones have always had an acute sense of the importance of precise measurement (small errors could mean big losses). Discrepancies among multiple carats were undesirable. In the 1870s some jewellers suggested the use of a karat of exactly 205 mg, but this solution had the risk of interfering with numerous national metrological laws that forbade the use of non-metric units in any commercial transaction.[29]

The solution to this quandary was proposed in the 1890s by the renowned mineralogist and vice president of Tiffany and Co., George Kunz. Due to his trade, Kunz was interested in reducing the transaction costs in the commerce of gemstones; but he was also akin to the idea of international metrication – Kunz was the first president of the American Metric Association (an influential pro-metric group founded in 1916). Kunz fruitfully mixed his knowledge and interests on both metrology and precious stones, and invented the so called 'metric carat', which ultimately helped to standardize and simplify measurement in that field. The metric carat was defined as exactly 200 mg.[30] This gave to the carat a perfect metric equivalency and a decimal notation. The scheme proved so effective that today the 'metric carat' has simply become the 'carat'.

This solution achieved several goals at once. It preserved the old name of the unit, which was familiar to jewellers and their clients; it offered a magnitude of the unit that was very close to the old carats, avowing a radical change that could be confusing and prone to be rejected by users; and as the 'metric carat' is an exact aliquot part of the gram, it synchronized the carat to the metric system. Probably a more logical solution would have been to drop the carat all together and simply employ the metric units in the commerce of precious stones. But what prevailed was a compromise between continuity and reform which was very quickly implemented among a reduced group of professionals.

When Tradition Cannot Be Demolished: The Failure to Revolutionize Units of Time

For centuries the *second* simply designated the fraction 1/86,400 of a day. This, of course, was the result of dividing the day into twenty-four hours, the hour into sixty minutes and the minute into sixty seconds. In this scheme the *day* was one of the centrepieces of time measurement – which had several variations in its definition depending on the epoch or culture, but all of them were related to the rotation of the Earth around its axis.

This basic understanding of what a second was was for the most part considered unproblematic until the 1950s, when astronomers became aware that the day was getting longer 'by about 1.7 milliseconds per century and that from year to year it could vary by the same amount'.[31] International conferences were held to fix this issue, which resulted in a proposal to redefine the second as a 'fraction of the tropical 1900', instead of a fraction of the mean solar day (which became unreliable due to irregularities in the rotation of the Earth). Thus in 1956 the International Committee for Weights and Measures adopted a new definition: 'The second is the fraction 1/31,556,952.9747 of the tropical year for 1900 January 0 at 12 hours ephemeris time'.[32] This agreement did not last long. Atomic clocks proved to offer greater precision than astronomical observations[33] and in 1967 the second was redefined again, now as 'the duration of 9,192,631,770 periods of the radiation corresponding to the transition between the two hyperfine levels of the ground state of the caesium 133 atom'.[34]

These definitions are practically as exact as the human intellect can manage at this point in history. But these modern definitions, as important as they have been to obtain greater precision and certitude, have not made any visible change in the reckoning of time for lay people (contrary to what happened with the metric system). We have not created a new second, we only updated the old one.

Today the second cannot be defined anymore as 1/60 of a minute; but it is significant that the opposite is still true: the minute is equal to sixty seconds (and sixty minutes still make an hour). People keep organizing their daily activities according to a time grid that would be familiar to persons living three centuries ago. We can see the relevance of this more clearly if we consider what changes in the units of time measurement have *not* been made.

The French revolutionaries in the 1790s aimed for 'the decimalization of everything measured or metered'.[35] Legislators and members of the Academy of Sciences took very seriously the idea that *all* divisions should be from tens into tens. As part of a plan for a thorough restructuration of all methods of measurement they redesigned customary weights and measures, the circumference, money and time units to fit into the decimal framework.[36]

Alongside the creation of the metric system they aimed for the creation of a new calendar and time-reckoning system: the Republican calendar and decimal time.[37] Instead of the irregular months of the Gregorian calendar, they opted for a more symmetrical partition, with thirty-day months, which were divided into three ten-day 'weeks' (called *décades*). The day was divided into ten parts (or hours), and each subsequent part into ten others.[38] The hundredth part of the hour was called *decimal minute*, and the hundredth part of the decimal minute was the *decimal second*. With this diagram all divisions of time, from the month to the second, were decimal.

Despite its elegance and its great similarities with the metric system, decimal time did not survive more than two years (it just lasted from November 1793 to April 1795), and was scarcely used in real life. Among the reasons adduced to suspend the proposal were the high costs of replacing clocks and watches, and popular confusion due to the novelty of the decimal units – not very convincing arguments considering that the metric system faced the same adversities, and was pushed through nonetheless. More persuasive was the argument that the old practices would continue 'due to the immense force of habit' and counting hours could not be policed by the state[39] (a recognition that at the time reform could to be effectively introduced without government coercion). And despite several tries to revive the idea of decimal time, this project has never gone beyond reduced groups of enthusiastic – though eccentric – scientists and aficionados.[40]

Conclusions

The failure to decimalize time illustrates the obduracy of units of measurement and the difficulty to make drastic alterations in 'folk' metrology without decisive political intervention. The impossibility to fully 'modernize' the second has forced metrologists to work under the straitjacket of traditional definitions of units of time measurement. Even the current atomic definition of the second of the SI is no more than a formalization of centuries-old understandings of time reckoning. The intertwined metrological understandings of experts and laypeople create defined limits to the transformation of units of measurement.

The case of the carat illustrates how reforming units of measurement can be relatively painless when it is circumscribed to the manipulation of understandings within the closed cognitive universe of a professional group. But things are different when the attempted alteration involves society at large – general stocks of knowledge are not as dynamic as the theoretical knowledge of experts.

To deeply renovate the metrological mores of a population much more than the agreement among specialists is needed. A robust social organization and conceptual machineries for the institutionalization, legitimation and maintenance of measurement systems are necessary. In the case of the metric system, that social organization had at its core the administrative power of centralized states (with their dual power to police and educate large populations); the conceptual machinery, on the other hand, was in the hands of scientists, who gave legitimacy to the system.

This sociological frame to understand measurement systems could be useful to assess the most recent debates to reform the SI: the redefinition of the kilogram. The current attempt to ground the kilogram on fixed values of fundamental constants of nature – freeing the SI from definitions of units based on artefacts – is a highly specialized reform. It has much to do with scientific communities. But this debate does not appear to be very relevant for the common-sense knowledge of billions of daily users who utilize the metric system in everyday life. These users, who are not at the forefront of innovation, are nevertheless part of the extensive network of institutions that conforms the metrological system at large; the metre, the litre and the kilogram were invented precisely with them in mind and they are ultimate reason why the metric system exists.

14 A MATTER OF SIZE DOES NOT MATTER: MATERIAL AND INSTITUTIONAL AGENCIES IN NANOTECHNOLOGY STANDARDIZATION

Sharon Ku and Frederick Klaessig

We have the metre rod in the archives. Do we also have an account of how the metre rod is to be compared...? Couldn't there be in the archives rules for using these rules-one used? Could this go on forever? ... we might put into the archives just one ... paradigm ...

Ludwig Wittgenstein[1]

Introduction

Standards used to be viewed as servants for science and technology innovation. Their mundane and routine characters make standardization less appealing to R&D, despite its crucial role in supporting the operation of technoscientific systems. However, the growth of global trade and the promotion of 'translational research' since the 1990s has fundamentally changed the role of standards in advanced technologies from 'unobtrusive servants' to 'faceless masters'.[2] They have, for instance, been promoted as 'soft rules' for regulating emerging technologies,[3] or been viewed as a tool for facilitating knowledge propagation from lab to market.[4]

While scientists and policymakers tend to view standardization and standard making as an objective solution for scientific controversies and technology regulation, science studies in recent decades have documented the constructive nature of standards and standard making. Historians of science examine standards qua standards, uncovering the deeply rooted historical, national, cultural and economic values embedded in objectivity,[5] uniformity and universality,[6] quantification[7] and precision.[8]

The socially constructed character only captures a part of standardization. For something to be called as a standard, it needs both contextualization and decontextualization. Bruno Latour portrays standardization as the 'center of calculation', a collection of 'immutable mobiles' such as writings and inscriptions, documents that enable the translation of interests and the ability to utilize the knowledge produced in various contexts.[9] Following this line, Joseph O'Connell further elaborates standards as 'a distributed collective connected by continually renewed structured relations of exchange and authority'.[10] Using metrology as case studies, he demonstrates that standard uniformity is a process of 'bottom-up' networking and extension, through the construction of 'traceability', establishing the authority of a particular representative, circulating it and assuring that comparisons are made to it. Timmermans and Berg argue that universality is always 'local universality' which depends on how standards manage the tension involved in transforming work practices, while simultaneously being grounded in those practices.[11] The role of standards as a negotiator is also argued by Joan Fujimura who proposes 'standardized package' as an assemblage of boundary objects with standard methods which then could be used for achieving agreement and managing collective action across social worlds to produce relatively (and temporarily) stable 'facts'.[12]

The brief literature review illustrates the complex, strong entanglement between science, technology and politics in standardization, summarized as the following two points: (1) The attribution of standards' objectivity, universality and uniformity shifts from scientific power to social and political authorities; and (2) Once the process of social construction is over, standards become powerful non-human actors that can be recruited to stabilize scientific knowledge production. However, we argue that these two arguments, both theoretically and empirically, do not fully capture standardization in contemporary science and technology development for the following reasons. First, recent studies in the IT, nanotechnology and biotechnology industries have shown that standard making in these advanced research fields has been integrated into global R&D competition and commercialization partnership building.[13] In such contexts, it is not just technical standards that need to be constructed, but also the national boundaries/interests in the international setting, and the structure, memberships and power relationships among the members within standards committees. The political uncertainties and contingencies in establishing standard institutions are no less than those in producing technical standards. These dual uncertainties indicate that it is insufficient to simply attribute standards' agency to institutional authority, viewing standards as 'centers of calculation' or as connections between local and remote 'authority', because these political powers, as well as the technical standards they endorse, are co-constructed by science and technology stakeholders.

The co-production of standards and standards institutions further points out a theoretical problem in STS: what do the social, political and cultural factors refer to in standard making? Scholars in science studies seem to focus more on epistemological questions about the values and contents of knowledge than the macro-institutional factors that address the political economy surrounding standardization.[14] Organizational sociology and political science take the institutional approach to study the organizational and international politics that shapes the production of standards,[15] yet there is less discussion connecting the macro-scale findings to micro-scale standard making and practice. However, our witnessing of and participation in various standard making processes shows that both local contingencies and structural institutional factors are equally important to understand standard making in emerging technologies,[16] as standardization is not just local consensus or collective action within scientific laboratories or disciplines, but also organizational behaviour, economic calculation and political tactics confined within formal, structured institutional settings. Concepts such as 'local universality', 'centre of calculation' or 'boundary object' cannot fully capture the institutional aspects of standardization. We need systematic frameworks integrating laboratory practices that produced the materiality of standards with organizational behaviour that endorsed the credibility of standards to elaborate the constructivist's view on standardization.

The connection between STS, political science and organizational sociology is particularly important to answer a critical but so far unresolved question in standardization studies: agency. Is the final outcome of standardization – collective rule following – dictated by material agency which represents the discipline of human beings through standard objects, or institutional agency which represents human beings' and institutions' control of objects? How to explain consensus reaching and collective action under the dual uncertainties from science and socio-politics and the co-production of scientific and socio-political authority? These questions require epistemological and ontological explanations; yet they have to go beyond being framed as just true/false or successful/failed claims, but a thorough socio-political issue that requires analyses at the institutional level. Most importantly, we argue that studying standards' agency is not just an STS research question, but also a critical issue for public policy. Given the reliance on standards in regulatory decision making, investigating the problem of agency also offers insight into the social responsibility of science and technology – who determined what the standards serving science and technology regulation were to be? We need more powerful conceptual tools and empirical data to challenge the social justice found in standard making: why can such a constructed 'universality' undermine other social or environmental fairness factors and be used as the reference for regulatory evidence?[17] Does the real contest in standardization rest on universality, or something else?

The interplay between material and organizational structures are key elements to understand the agency of standards. It is not possible to tackle this question fully at both the theoretical and empirical level in such a short essay. Therefore in this paper, we plan to summarize our research and participation in nanotechnology standardization as a starting point to initiate more dialogue among STS, organizational sociology and science policy studies in contemporary technoscience standardization. Our empirical data is drawn from the production and application of the first nanotechnology size standard, Gold Nanoparticle Reference Material (gold RM), certified by the US National Institute of Standards and Technology (NIST).[18] We use 'the biographies of scientific object'[19] as the analytical framework to capture the material and institutional agency constructed at different stages of the gold RM – how it was characterized, certified in NIST and later applied to the Inter-Laboratory Study (ILS) organized by ASTM International for the validation of nanomaterial characterization protocols. Tracing this complete life cycle, we propose the following four theses as our answer to the question of the agency of nanotechnology standards:

Thesis 1: Atomic precision does not guarantee nanoscale standardization

Thesis 2: Standard objects are not the source, but the outcome of disciplined action in a structured organizational setting.

Thesis 3: There is no predetermined authority to dictate the obedience to standards, either from the standard object or standard institutes.

Thesis 4: Following a standard is a communal practice that cannot be dictated by any single object or institution. The material and institutional agencies of standardization come from the constant re-iteration by the standard users and members of the organization.

Theses (1) and (2) are demonstrated by the production of the NIST-certified gold RM, while theses (3) and (4) will be elaborated in the case study of ASTM ILS.

NIST-Certified Gold Nanoparticle Reference Material

Thesis 1: Atomic precision does not guarantee nanoscale standardization

Nanotechnology is a rising field that has caught political, scientific and economic attention since the announcement of the US National Nanotechnology Initiative in 2000. It is portrayed as a technology that can manipulate matters on an atomic, molecular and supramolecular scale, from 1 to approximately 100 nanometres.[20] The canonical history of nanotechnology normally starts from the invention of the scanning tunnelling microscope in the 1980s which made 'seeing' and manipulating atoms possible. With an atomically precise tool, exact measurements and control at the nanoscale, which is ten times larger than the atomic scale, can be obtained to characterize and design novel material properties and functionality for environmental and biomedical application.[21]

This linear framework has been mobilized by the US government to fund the National Nanotechnology Initiative for close to 20 billion USD since 2001. However, the journey for nanotechnology to travel from angstrom physics to other applied fields turned out to cross the notorious 'valley of death' in the process of translation.[22] For instance, the main funding agency of nanomedicine, the National Cancer Institute (NCI), had noticed that many laboratories which claim to use the same nanoparticles as drug carriers reported inconsistent particle size. The poor reproducibility of this very basic parameter not only increases uncertainties in any size-dependent scientific claims, but also raises concerns from the US Food and Drug Administration (FDA) about the quality of manufacture and measurement. Obviously, instruments capable of characterizing atoms apparently fail to capture objects consistently at the nanoscale.

The solution NCI proposed was to produce a standard to calibrate various measurement protocols. It decided to work with NIST and the FDA, bringing the metrological and regulatory concerns into the development of nanodrug characterization. A new facility – Nanotechnology Characterization Laboratory (NCL) – was established in 2005 to represent the interests of the nanomedicine community; the FDA-NCI-NIST Memorandum of Understanding was signed in 2006 to facilitate the interagency collaboration.

What the NCI expected from NIST was a standard object with a quantified size that could be put side-by-side with the measured samples as to calibrate size measurements from different research sites. NCI believed that making this one-dimensional size standard was an easy and straightforward task for the NIST metrologists who have the expertise and tools for producing much more sophisticated standards. Nevertheless, this optimistic thinking was overwhelmingly discouraged by NIST metrologists who argued that 'for NCI, size is just one number, but for NIST, a standard size claim requires thousands of measurements, statistical calibration and uncertainty analysis'.[23] The interpretational flexibility also occurred in defining the meaning of 'standard'. The NCL biologists demanded a standard stable in and compatible with biological systems; yet NIST metrologists challenged such a presumption, arguing that 'biology is an open system full of unexpected variables. Any trustworthy standard cannot be developed within this unstable environment'.[24]

The prolonged debates about the philosophy of the size standard lasted for almost two years. Three NIST liaisons resigned from the position; none of them could avoid these circular discussions and move the collaboration forward. Yet, these technical debates were more than fighting about who was right or wrong. They were in fact strategies each agency used to establish its institutional identity in the inter-agency collaboration. Before this collaboration started, there were already internal doubts from NIST employees, questioning the impacts of partnering with the NCI. Compared with NIST, the NCI is a much bigger, much wealthier organization. NCI's intramural budget is already several times larger than the whole NIST budget. Given the organizational asymmetry, NIST liaisons were extremely conscious of safeguarding NIST's autonomy while pursuing the collaboration. The strategy taken was to mobilize the term 'standard' as a tool to defend NIST's critical role and establish NIST's leadership identity, as one of the liaisons stated, 'at NIST the word "standard" means something which requires very strict definition in metrology … what NCL has in mind is an assay protocol, a quantitative measurement, not a standard'.[25]

The NCL had its own identity issue. As a government-owned, contractor-operated research facility, the mission of NCL is to provide research requested from its boss the NCI, its partner the FDA and its clients the profit-oriented drug developers. Yet being a federal contractor working in this highly risky field with no permanent job security, the NCL had to be strategic and practical about defining the objectives of standardization, 'not to develop a standard with an infinite decimal accuracy, but "standards" to assist our clients in getting their nanoparticles into clinical trials', said by the NCL director.[26]

A nanosize standard, a scientific object initially associated with the absolute atomic precision, turned out to be mostly a political tool, mobilized by the agencies to protect their interests and autonomy in partnership building. The measurement of size, the meaning of standard and the identity of the institutes, form 'the standardizer's regress' that prevents any standard measurement to be obtained solely by technical means.[27] The interpretational flexibility of size standards at the organizational level indicates that breaking the regress requires organizational reformation.[28]

Thesis 2: Standard objects are not the source, but the outcome of disciplined action in a structured organizational setting

These two agencies needed a match maker to repair the broken relationship between the 'stubborn' NIST metrologists who always wanted to be in control, and the NCL biologists who believed that the varieties and uncertainties in the real world cannot be simply depicted by rigid metrology. Debra Kaiser, chief of the ceramic division at NIST, was appointed to be the person leading the project. Unlike her predecessors from the Physical Laboratory who tended to define standard as fundamental metrology, Kaiser had participated in several projects developing standard testing material for the ceramic industry. These practical experiences allowed her to define a standard in this case as a 'product' requested by a client, the NCI. 'NIST had to treat NCI like a client; we work with our client to produce standards based on clients' interests and the intended use of the standard'.[29]

Redefining the NIST–NCI partnership as a customer relationship rather than a political one reduced the competition between the two agencies and re-focused the NIST bureaucracy on the production of standard reference material for this inter-agency collaboration. Every standard object NIST produces has to follow the steps according to the Office of Reference Material, illustrated in Figure 14.1: first it requires a pre-production survey to assess the market need, cost, technical and statistical demand. Entering the production phase, the responsible laboratory must perform measurements according to the advice given by the statistical division, which conducts the statistical analysis to determine the mean value and uncertainty. After all the numerical data are produced, the process moves to the post-production phase, which includes the tasks of documentation review, packaging and labelling, pricing and marketing by the service division.[30] That said, a NIST certified standard cannot be accomplished in a single lab by an isolated individual; instead, the measurement task is divided into various steps that rely on coordination among difference offices within the NIST bureaucracy.

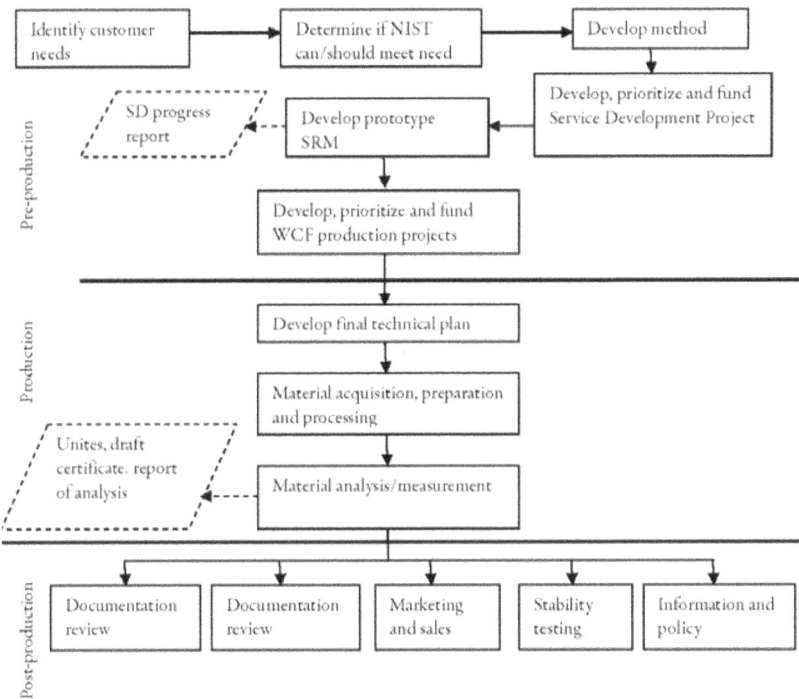

Figure 14.1: Illustration of the procedure of NIST SRM production

The NCL requested that this standard to be used in calibrating various instruments, as the FDA demands not only one measurement, but rather a cascade of measurements to validate the quality of nanomedicine. The instruments NCL suggested were located at different research laboratories in NIST, raising another level of coordination the project manager had to resolve. Furthermore, since instruments vary in operating principles, they lead to different numerical expressions which required NIST to turn this size standard into a multiple value standard in order to calibrate multiple instruments. Producing a multi-value standard brought new challenges to NIST, as in the past, all NIST size standards were for a single value targeting a specific instrument calibration. Putting different numbers together violates the intuitive view of size as 'one number', and caused confusion in the nanotechnology community about which measurement is more accurate, and which number is closer to the true value of a particle's size.

All the expectations and constrains on the gold RM suggest that the production of this standard object is not solely about measurement, but a problem of management. To turn the NIST bureaucratic machine from a stubborn opponent into an efficient alliance for this collaboration, Kaiser created a cross-institute working group to reorganize the NIST–NCL interactions by imposing a series of managerial plans. Instead of relying on the existing division of labour in NIST, she used the names of nanoparticles to reclassify experts, resources and information involved in this working group. This reorganization temporarily liberates NIST staff from their daily bureaucratic duties when working across agencies, therefore allowing the working group to became a trading zone for the NCL and NIST scientists to share samples, instruments and data. In order to avoid controversies among different measurements, which could potentially ruin the credibility of the standard, Kaiser mobilized the NIST bureaucratic structure for data collection. Researchers only attended the group meeting specific to their assigned measurement task and reported data only to the group leader. Table 14.1 is the table used in the NIST official report that contains size information of the gold RM. Seven numbers obtained from seven instruments were juxtaposed in a particular order without mentioning the correlations and comparisons among these different representations. This information order was established through the 'bureaucratic indifference': the bureaucratic mentality of 'mind-your -own-business' offers the infrastructure to create an information order for harmonizing these 'multiple truths' generated by different experimental systems.[31]

Table 14.1: Size as a multi-value parameter in the case of gold RM

Technique	Analyte form	Particle size (nm)
Atomic force microscopy	Dry, deposited on substrate	8.5 ± 0.3
Scanning electron microscopy	Dry, deposited on substrate	9.9 ± 0.1
Transmission electron microscopy	Dry, deposited on substrate	8.9 ± 0.1
Differential mobility analysis	Dry, aerosol	11.3 ± 0.1
Dynamic light scattering	Liquid suspension	13.5 ± 0.1
Small-angle X-ray scattering	Liquid suspension	9.1 ± 1.8

NIST bureaucracy, once successfully mobilized, becomes a powerful alliance in stabilizing scientific 'facts'. It not only provided the social order to prevent potential scientific controversies at the research frontier; it also guaranteed the robustness of the gold RM as a mundane calibration device for routine laboratory work. The gold RM was released with guidelines instructing users of the proper procedures for using the standard. Even a trivial action like opening the bottle has a written instruction that one should open the ampoule by 'applying moderate pressure with one's thumb to snap off the nipple'.[32] These rules protect the credibility of the gold RM through their implementation within the bureaucracy. NIST SRM developers used to receive phone calls from users doubting the quality of the SRM, due to their frustrations of not being able to obtain the same standard values NIST suggests. Yet most of the complaints ceased after the agency instituted a formal, detailed, time-consuming problem shooting procedure, to make sure that the instructions of use have been carefully followed. In other words, the agency of discipline does not directly come from the standard object, but from the rigid division of labour within NIST to reduce both scientific uncertainties and social contingencies. To question the credibility of the gold RM, one needs to be prepared to fight with the whole NIST bureaucracy.

In Jan 2008, three gold reference materials with nominal sizes 10 nm, 30 nm and 60 nm were finally released by NIST (Figure 14.2a). It took three and a half years for NIST and NCL to make the size standard, because the whole process is not just a production of a material object, but a co-production of the material structure and its corresponding organizational infrastructure. This organizational infrastructure is the first thing shown on the Report of Investigation, the official certification for the gold RM (Figure 14.2b). All the participating divisions and staff are listed on the first page. Gold RM, once declared by the NIST, is no longer just a technical object, but a bureaucratic document which ties precision closely with bureaucracy. Note that the NIST bureaucracy did not act as a predominant authority dictating the gold RM production; quite the opposite, it is the local adjustments of NIST employees and the mobilization made by the project leader that active the function of bureaucracy in standard making. When scientists perform and report the measurement, they at the same time are exercising rule following in NIST bureaucracy which declares the NIST bureaucratic authority. The organizational structure and the material structure of the gold RM therefore mutually resemble each other, and it is this mutual resemblance that results in the gold RM's technical and social robustness.

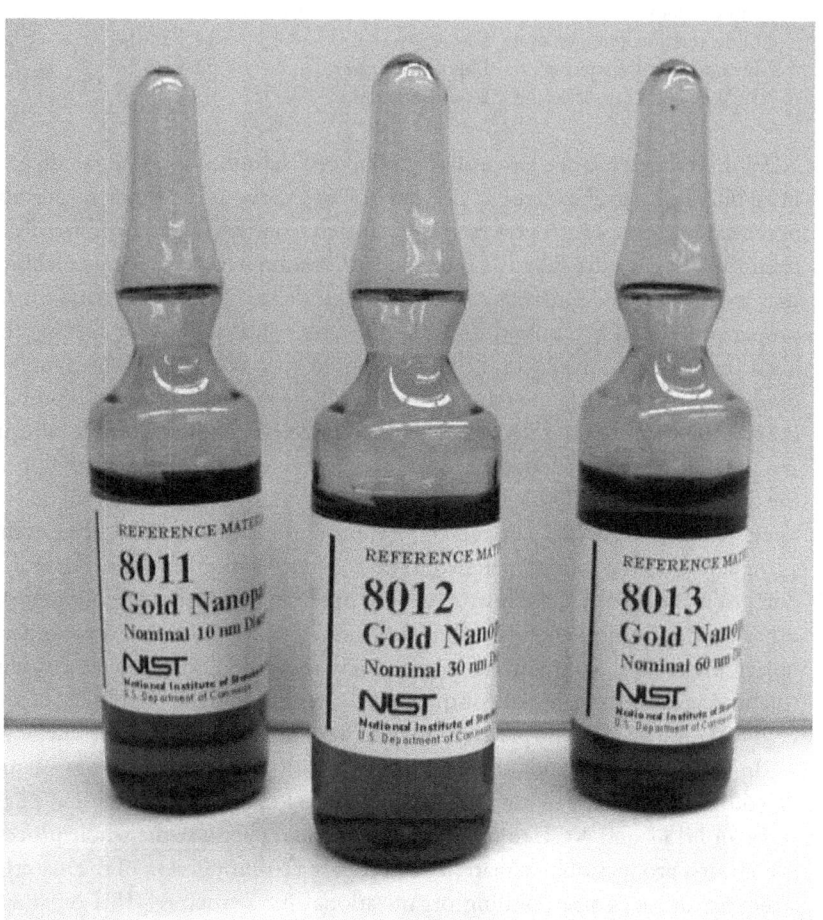

Figure 14.2a: NIST gold RM

National Institute of Standards & Technology

Report of Investigation

Reference Material 8011

Gold Nanoparticles, Nominal 10 nm Diameter

This Reference Material (RM) is intended primarily to evaluate and qualify methodology and/or instrument performance related to the physical/dimensional characterization of nanoscale particles used in pre-clinical biomedical research. The RM may also be useful in the development and evaluation of in vitro assays designed to assess the biological response (e.g., cytotoxity, hemolysis) of nanomaterials, and for use in interlaboratory test comparisons. RM 8011 consists of nominally 5 mL of citrate-stabilized Au nanoparticles in an aqueous suspension, supplied in hermetically sealed pre-scored glass ampoules sterilized by gamma irradiation. A unit consists of two 5 mL ampoules. The suspension contains primary particles (monomers) and a small percentage of clusters of primary particles.

Expiration of Material: The reference values for RM 8011 are valid, within the measurement uncertainties specified, until **31 December 2012**, provided the RM is handled in accordance with the instructions given in this report (see "Instructions for Use"). However, the size distribution may be altered and the RM invalidated if the material is contaminated or handled improperly.

Maintenance of Reference Values: NIST will monitor representative samples from this RM lot over the period of its validity. If substantive changes occur that affect the reference values before the expiration date, NIST will notify the purchaser. Registration (see attached sheet) will facilitate notification.

The overall technical coordination for material procurement, processing and measurement activities was conducted by V.A. Hackley and J.F. Kelly of the NIST Ceramics Division.

Reference and informational value measurements were performed at NIST by the following: NIST Analytical Chemistry Division: T.A. Butler, R. Case, K.W. Pratt, L.C. Sander and M.R. Winchester. NIST Ceramics Division: A.J. Allen, T.J. Cho, J. Grobelny, V.A. Hackley, D.-I. Kim and P. Namboodiri. NIST Metallurgy Division: J.E. Bonevich and A.J. Shapiro. NIST Polymers Division: M.L. Becker, D.L. Ho, A. Karim and B.M. Vogel. NIST Precision Engineering Division: B. Ming and A.E. Vladár. NIST Process Measurements Division: L.F. Pease III, M.J. Tarlov, D.H. Tsai, M.R. Zachariah and R.A. Zangmeister.

Statistical consultation on measurement design and analysis of the reference value data were performed by A.I. Avilés of the NIST Statistical Engineering Division.

Additional technical and coordination aspects were provided by the following: R.F. Cook, W.K. Haller and D.L. Kaiser of the NIST Ceramics Division.

Support aspects involved in the preparation and issuance of this RM were coordinated through the NIST Measurement Services Division.

RM 8011 was developed at the request of the National Cancer Institute (NCI). Development and production costs were subsidized by NCI.

Debra L. Kaiser, Chief
Ceramics Division

Gaithersburg, MD 20899
Report Issue Date: 13 December 2007

Robert L. Watters, Jr., Chief
Measurement Services Division

Figure 14.2b: Report of investigation

Gold RM as Gold Standard? ASTM Inter-Laboratory Studies

The release of the gold RM is not the end of the story. To turn the gold RM into a truly 'Gold Standard', this object has to depart from the NIST, traveling to users' labs to gain the acceptance from the nanoscience community. One might feel confused about why this step is necessary, given the gold RM has been certified by NIST, a national standard institute which should possess the authority in standard declaration. This question constitutes the second part of the paper: the role of the state in standardization. We use the case study of the gold RMs application on the ASTM Inter-laboratory studies to discuss this subject.

Thesis 3: There is no predetermined authority to dictate the obedience of standards, either from the standard object or standard institutes

The role of the state in contemporary standardization has to be situated within the complicated public–private partnership involving government and industry, where the state actually plays multiple roles as the sovereign of regulation, the patron of investment or the user of industrial standards. Taking the US as an example, the American standard system reflects a strong cultural and political bias in favour of market solutions.[33] American standard development organization (SDOs) have a long tradition of keeping government and its mandated role at arm's length. Without government control or any other central monitoring and coordinating agent, the US standard system is characterized by extreme pluralism. Most SDOs are voluntarily organized by firms and technical practitioners to solve practical issues in the product development processes. Each standard organization has its own consensus-reaching mechanism that is used to help members reach agreement in the development of 'voluntary consensus standard'. The neo-liberalism view on standardization not only offers industry enough freedom to make their own standards, but also requires the federal government to use and conform to voluntary consensus standard developed by private SDOs.[34]

The American Society for Testing and Materials International (ASTM International) is the largest non-governmental SDO in the US dedicated to voluntary consensus standard making. It has over 10,000 members from industry, government and academia, constituting over 100 technology committees, including nanotechnology E56 committee (established in 2005).[35] Each committee organizes regular consensus meetings to evaluate standard activities initiated by members. To organize diverse interests from a wide range of stakeholders and guarantee the efficiency of standard making, ASTM imposes a series of social infrastructures to guide members' collective action, requesting each committee to operate under the ASTM administrative guidelines. For any local measurement (including the results from government laboratories) to become an ASTM-certified standard, it must pass through a formal, structured and hierarchical procedure. First, protocol developers need to find a way to solicit comments on their protocols. The revised drafts are then submitted to the committee as work items for members' review. Consensus ballots will be used to determine whether a work item can be certified as an ASTM International standard. Individual members associated with the committee are obligated to vote on the proposed documents and whose votes have the same weight. Since the ASTM insists on an equal footing of small, medium and large sized interest parties which might have numerous representatives on a committee, it allows any negative vote from any member to stop a ballot item from moving forward until it is resolved. Under this mechanism, federal agencies have very limited power in dictating the procedure and products of standardization. The state, like other interest parties, need to express its support of / objection to standardization by following the ASTM procedure. Therefore, even the gold RM that has been carefully measured and formally certified by NIST is still not considered as a voluntary consensus standard endorsed by the nanobio industry.

Parallel to the development of the gold RM, the NCL had developed over thirty protocols for both physicochemical and biological nanomaterial characterization protocols. These protocols, just like the gold RM, though repeatedly tested within the government-owned laboratory, have not fully established its credibility within nanobio community. In order to turn this federally funded research into consensus standards, the NCL decided to submit three basic but crucial testing methods – particle size, cytotoxicity and hemolysis– all with the gold RM as the standard testing material, to the ASTM International E56 Nanotechnology Committee, getting them certified through the procedure described above.[36]

To solicit comments on these protocols and develop a precision and bias statement for each test protocol, the E56 Committee recommended that the NCL and NIST conduct inter-laboratory studies (ILS) using NCL standard assays and NIST gold RM as testing materials. The ILS is a unique platform used by the ASTM International to solicit users' comments for validating the quality of standards and test methods. As an institution developing standards for industrial purposes, ASTM International honours 'real-world precision' which defines the true value as a consensus value derived from the experimental results produced by the ILS participants.[37] To obtain this real world precision, ASTM International recruits participants to conduct round robin testing using the same testing materials and same written protocol; each participant is required to provide standard deviation analysis and quality control in a data report to demonstrate that the test methods are repeatable and reproducible.

Among the three testing methods, hemolysis assay was considered, from the NCL's perspective, a simple and easy enough test. This assay quantifies nanoparticle-induced hemolytic properties to characterize the toxicity of nanodrugs. The gold RM was incubated in human blood sample; if the blood cells are damaged by the gold, hemoglobin will be released and converted to red-coloured cyanmethemoglobin measured by spectrophotometry, a turn-key measurement.

The ILS started in 2007. The NCL distributed the testing toolkit, including the hemolysis protocol, blood samples and two gold RMs (30 and 60 nm) to nine volunteer laboratories. After numerous emails, phone communications with the participants and months of waiting, the NCL scientists finally received the results, which were, according to the ILS coordinator, 'extremely frustrating and out of our expectation after spending so much time preparing the study'.[38] Nine labs received the test kit, yet only seven labs responded to be in the final report. For the two that did not submit data, one experienced sample contamination; the other did not stay in contact. Among the seven responding labs, only two successfully followed the protocol 'as written' and submitted completed datasets. Four submitted incomplete datasets; one lab missed the testing plate that made data analysis impossible. There were insufficient results for a statistically meaningful precision statement.[39]

The unexpected outcome raised concerns within the NCL about the quality of the protocol and the gold RM. However, the NCL scientists later found that it was not the lack of technical details that prevented participants from following the protocol, but rather the hidden cultural values embedded in standard following. Due to the novelty of the subject, the nanotechnology ILS participants included several university labs. Professors coming from research-oriented universities are usually more interested in pursuing 'novel experiments' to generate data that can be published in scientific journals, rather than spending time on following someone else's protocols to validate precision statements. Therefore, this 'uncreative' work was often assigned to students who do not have GLP (good laboratory practice) training to understand and ensure the 'uniformity, consistency, reliability, reproducibility, quality, and integrity of tests'.[40] The lack of GLP training, the incentive, infrastructures that support standardization and the culture of appreciating the value of standards inevitably led to problems of understanding, interpreting and performing the measurement protocol. For instance, some labs did not complete the test because the fresh blood samples were not promptly picked up when the package arrived. Some lost or contaminated the testing plates; others did not follow the written text line by line but modified them according to their own practice. Even for those who claimed to be thorough and accurate, there were still alternative interpretations of the protocol's text. For example, 'mixing A reagent with B solution': does it mean adding A to B or B to A?

The interpretational of the ILS result also generated debate with the committee. Physical scientists participating in the balloting process were surprised to hear that 20 per cent, or even greater, variability in biological assay results is considered acceptable. The physical scientists attributed this high level of uncertainty to the lack of careful design of experimental controls in quantifying the toxicity generated by nanoparticles. However, the biologist who designed the protocol argued that the precision statement ASTM International required reflects its engineering mindset that everything ought to be, and can be, in control, and completely ignores the reality that most biological components like cells are subject to adaption and change over time. This further suggests that the precision statement for the ILS is value laden; the values of precision are determined by disciplinary conventions and paradigms of knowledge production.

The ILS result shows that the NIST gold RM, though certified by the national standard laboratory, does not have the predominant authority to dictate obedience to the standard in ASTM International's ILS. Once the standard object and its testing protocol depart from the original point of NIST certification, they no longer act as harmonizing devices to discipline human action. Instead, they become irritating objects that perturb the social relationships among heterogeneous nanotechnology stakeholders. Tensions in interdisciplinary and translational research are amplified through these standard objects to test the foundations for coordination and trust within the broader nanotechnology community. However, instead of viewing the gold RM as a 'bad standard' that cannot be followed, we argue that that these irritations constitute the ontology of standardization. They are inevitable outcomes of the interdisciplinary nature and knowledge pluralism embedded in nanotechnology. They also reflect the market orientation and decentralization of US standardization policy. These irritations are the critical components of standardization, as they 'reify' abstract cultural and social differences into concrete and negotiable work items such as experimental controls, protocol texts or labour training. The gold RM and the ILS further confine these irritations in an ordered ASTM International infrastructure which offers mechanisms for negotiation. Stakeholders involved in standard making have to express their opinions by accepting and knowing how to follow the organizational rules, i.e., becoming a member of the organization and participating the balloting process.

Thesis 4: Following a standard is a communal practice. The material and institutional agencies of standardization come from the constant reiteration by the standard users and members of the organization.

The poor reproducibility of the gold RM characterization protocol indicates that standard following is not dictated by any precise object or authoritative institution. In fact, both 'precision' and institutional authority should be viewed as the outcome of collective rule following. This can be further elaborated through

Wittgenstein's thesis on rule following and SSK's philosophical basis on meaning finitism:[41] Standards do not guarantee how they can be followed. The future application of standards is open, and collective standard following relies on constant reinterpretation, reaffirmation and readjustment based on the exiting consensus within the community. Scientists within the NCL know how to follow the protocol, because there is a collective understanding about the value of standard measurement in producing FDA-recognized results, with the organizational culture that encourages reproducibility, stability and quality control. In addition, members within the NCL have the embodied knowledge built through their daily lab practice to understand the tacit aspect of the protocol, which is difficult to spell out completely in texts. However, non-NCL members do not share these common values and beliefs. As a result, even written protocols and a certified testing material left open 'interpretational flexibility' in standard following.

The ASTM E56 committee, after a long process of reviewing the ILS and communicating with the NCL scientists about the improvement with the protocols, finally accepted that the NCL hemolysis assay, though not perfect, was probably the best available protocol to measure hymoletic properties of nanomaterials.[42] The committee members concluded that although the testing phase of these studies was ruled 'inconclusive', they had collected valuable information about the planning and management of future protocols and supplies, 'dealing with everything from the difficulty of transporting samples during a New England winter, to how many decimal points are realistic to use in nanoparticle measurement'.[43] Therefore, most of suggestions made did not directly target the measurement per se, but addressed increasing the collective understanding and practice towards the tacit knowledge embedded in protocol following, such as sample preparation, through organizing protocol-training workshops before the round robin test.

The ASTM ILS illustrates the social foundation of precision and standardization.

To transform NIST-certified gold RM into truly Gold Standard followed by standard users, and the NCL standard protocol to be voluntary consensus of the nanobio community, measurement tools with decimal precision is not the solution; rather, it is the creation of 'the community' – the shared values, infrastructures and beliefs – that grants 'precision' and 'standards' the social and technical robustness. Following a standard should therefore be viewed as a communal practice, as the co-chair of the ASTM E56.02 subcommittee pointed out, '[The ILS] It's like trying to design a diet program for the entire population – it just isn't going to work. You need to tailor your approach to each individual'.[44] An effective standard demands the users' collective action to declare its technical superiority; yet such collective action in a voluntary consensus standardization process cannot be mobilized by centralized institutional power, but rather by members communicating, negotiating and committing to the community. Building the appropriate social base and social infrastructure to support collective rule following, we argue, is the essence of standardization.

Conclusion

In this paper we propose four theses to offer a sociological account of standardization. Through the biography of the gold RM, we demonstrate that the agency of standardization does not come from the non-human objects per se, but from the constant re-enforcement of shared beliefs among members of the standard community. A matter of size can only matter if there is a user community with the understanding, skills and trust to follow the measurement protocols. It is the collective rule following that grants the gold RM the agency of being a standard for calibration.

Standardization therefore is a communal process that relies on a robust community as the social base to support collective rule following. Such social infrastructure building, however, has not been fully recognized by the current nanotechnology science and policy communities. We thus conclude this paper with two policy suggestions:

(1) Understanding and legitimating the values of precision

The social and cultural values embedded in the gold RM ILS precision statement challenges the current usage of 'atomic precision' in promoting and legitimating the development of nanotechnology. Mobilized as a 'value-free' quality, the use and conceptualization of atomic precision ignores the reality where diverse social values and interests contest over the question about 'how precise is enough, by whose standard'? Precision thus is not the solution that nanotechnology can offer; rather, it is something that desperately needs social explanation and justification.

The problematic usage has caused, and allows, unrealistic policy statements and questionable decision making based on the misunderstanding of precision. Developing a conceptual framework to recapture ontological and sociological aspects of technoscientific precision and standardization, we believe, is a crucial step to connect nanotechnology measurement and standardization to broader ethical and societal demands. Only by revealing the social values embedded in precision measurement can we see precision as an issue of public policy and be able to examine whose precision and which standards speak for the public goods.

(2) Understanding and incorporating social infrastructures into standardization policymaking and education

Standard following as a communal practice requires social bases. However, scientists, policymakers and standards organizations still consider the production of scientific standards to be a purely technical process that relies on scientific expertise to design and judge the outcome. Recent efforts initiated by the US NNI to build the Nanotechnology Knowledge Infrastructure to improve data sharing and standardization is one example.[45] The whitepaper, though mentioning the importance of community building, focuses overwhelmingly on creating the cyber-infrastructure and pays little attention to the social infrastructure, which, according to our empirical findings, is the 'bedrock' of technical standardization. The agency of harmonization and standardization does not come from non-humans; rather, it is our collective understanding, action and belief that activate non-human objects to serve as standards. We suggest that more interdisciplinary dialogue among standards production, policymaking and socio-philosophical studies of standardization should occur for down-to-earth nanotechnology research policymaking. This co-authored paper by an industrial standard developer and an STS standard researcher is an initial attempt to validate this proposal.

Acknowledgements

This material is based upon work sponsored by the National Institute of Standards and Technology Science Guest Research Program and the National Institutes of Health Stetten Fellowship Program. The authors wish to thank the NIST SRM development team and the Nanotechnology Characterization Laboratory at the Frederick National Laboratory for their support of the fieldwork. Any opinions, findings and conclusions or recommendations expressed in this material are those of the author and do not necessarily reflect the views of the above agencies.

15 MEASURING BY WHICH STANDARD? HOW PLURALITY CHALLENGES THE IDEAL OF EPISTEMIC SINGULARITY

Lara Huber

The fact that the cesium tube is currently declared the most authoritative representation of the time interval standard by national and international agreement is a social fact that merely recognizes the natural fact of its superior accuracy.

Joseph O'Connell[1]

Regarding strategies of measurement in particular, and issues of comparing scientific data in general, both endeavours are highly connected to the normative power of scientific norms, so-called standards. The common understanding of a standard for scientific purposes reads as follows: technical or rather methodological standards are instruments for stabilization and/or validation – given that they are established and acknowledged norms of a specific scientific practice or scientific purpose, such as measurement. The classical approach of standard setting would be to identify a certain epistemic goal (i.e., uniform and precise measurement), and hence to introduce a certain scientific norm (i.e., measurement scale) to allow for that very goal. The process of realizing the epistemic goal is therefore called standardization or rather normalization,[2] due to the fact that research data is evaluated on the basis of a given reference set of data and a standard gauge (normal data, normative sample, etc.) respectively. Hence, standards might be purely and primarily technical in their nature. With regard to scientists' practices, such norms more often than not have a function that transcends their mere technical nature, namely if this very norm exhibits a genuine epistemic function by validating scientific data.[3] Standards give rise to values, such as precision,[4] which in turn are seen to impact on scientific ends, such as objectivity.[5]

This paper reflects on the question to which extent standards express authority, for example with regard to the validation of scientific data. Here I would like to touch upon the issue of the evaluative and normative nature of scientific standardization. The first step of analysis, under the heading 'Rising Standards', is concerned with the question of how scientific norms evolve, or rather, of how they are introduced for scientific purposes. Secondly, the paper focuses on the epistemic rank of standards and critically assesses the ideal of epistemic singularity. The third and final step addresses challenges that arise out of contemporary research practices. Especially in data driven sciences (i.e., climate research; -omics research: proteomics, genomics, etc.), parallel regimes of standardized formats call for standardizing the use of standards by introducing standards of a higher order: so-called 'metastandards'.

Rising Standards

In general, scientists know exactly which norms or conventions to consult if they aim to properly validate their findings. Standards, i.e., guidelines or protocols, are rarely critiqued. This is because they are so urgently required: they give direction, organize scientific practice and allow for validation or even verification.[6] The workplace and practice of a scientist is guided by standard formats that impact on the design and the production of data, as much as on the assessment of research results. Standards are more often than not simply in place, mandatory prerequisites for meeting the very basic conditions of contemporary research (i.e., good clinical practice in FDA-regulated clinical trials). Concurrently, due to technical innovation and scientific progress, standards in a way remain objects of improvement, or even of displacement. A new standard may evolve within the realm of a given research practice ('ad hoc standard')[7] or rather, it may be introduced as a prerequisite of further research by a consortium with a predefined scientific purpose (i.e., the protocol for tissue sampling). It seems necessary to differentiate between the primary processes of the evolution of a standard format and further processes of implementation. Let us presuppose that we identify, within the realm of different working standards, a potential candidate for a scientific norm. Now, further processes of implementing this candidate as a standard gain momentum. In the following I differentiate between two types of implementing standards, which, nevertheless might be interdependent, or happen simultaneously. Both types illustrate the ratio of acceptance and acknowledgement: a peer group recognizes a format as a scientific standard,[8] given that it has been generated on the basis of scientific excellence itself.

(A) The first type of implementing a format as a scientific norm includes scientific practices that stabilize both its validity and its use with regard to a given epistemic goal. Hence, one could speak of implementing a standard format for scientific purposes through in *progress-stabilization*. There could be two reasons for in progress-stabilization. Ideally, scientists prefer a certain standard format (e.g., protocol) because of its extraordinary functional capacity to achieve a certain goal of standardization. It might also be the case that the given protocol is not that useful for this purpose, but the best one to fulfil at least a certain aspect or simply the only one at hand (pragmatic standard setting). The second reason relates to this pragmatic approach. Here, a standard format, regardless of its genuine functional capacity, is commonly referred to by scientists because it helps connect their research to cutting-edge technology. In psychological research, p. ex., critics examine to what extent the inclusion of imaging devices, such as fMRI, is a must for contemporary research into cognitive states.[9]

(B) The second type of implementing a standard format includes all endeavours to set standards for scientific purposes in an explicit way. Examples for explicit fixations are the creation of the International Vocabulary of Measurement (VIM) and regulations relating to good clinical practice and clinical trials by the US Food and Drug Administration. Two aspects of *explicit fixation* need to be kept in mind. First, strategies of implementing standard formats are more often than not based on an economic factor that helps finance further research. Or they could even create from scratch, as the example of bio-industry shows, a whole new marketplace that provides researchers with a wide spectrum of standard formats, ranging from analysis kits to purifying plasmids to genetically optimized living organisms. Apart from mere financial aspects, setting standards constitutes a significant factor for fostering scientific networking around a given standard format, especially if it is highly functional with regard to a given standardization purpose. Classic examples are metric prototypes, classification systems and nomenclatures, as much as reference data (i.e., Craig Venter Genome). Today, standard setting is only rarely the subject matter of a single expert in his field, but rather the result of a huge network of scientists forming large research consortia (i.e., World Weather Watch, Human Genome Project, Large Hadron Collider). Still, it is noticeable that there is often quite a strong relationship between a pre-existing authority in a given field of research, for instance a well-operating consortium that is highly acknowledged within its scientific community, and the authority of a standard format, and accordingly, its presumed epistemic rank within the very same field of research. The latter addresses priority issues; the question arises why a certain format that has been set as a standard is prioritised or rather neglected in the further process of implementation. Do we have to review the process of implementing a standard format?

Introducing the Ideal of Epistemic Singularity

First, let us confront a significant discrepancy between the very process of standard setting and the acknowledgement of a standard format, which could be understood as continued processes of implementing a format as a standard. Both issues, standard setting as well as implementing a format as a standard, directly relate to a key aspect of scientific validation, which I would like to call epistemic singularity of a standard (= scientific norm). According to this ideal, a standard is considered epistemologically singular, if no other scientific norm that is produced on the basis of scientific expertise carries an equal, or even higher efficacy, from an epistemological point of view (i.e., precision of a standard gauge). Hence, epistemic singularity refers to the fact that a scientific norm is singular in character, especially with regard to other scientific norms already in existence (or even yet to come), addressing the same phenomenon, i.e., temperature. Briefly, two criteria for the epistemic singularity of a standard coexist. The first criterion reads: a scientific norm is *singular in character*, if – and only if – just one standard exists or is scientifically approved for the assessment of the quantity and/or quality of a phenomenon in question. The second criterion reads: a scientific norm *performs highly from an epistemic point of view*, if – and only if – a given standard is exceptionally functional with regard to the assessment of the phenomena in question. Ideally, both criteria come together. Hence, while speaking of a given scientific norm as a standard, i.e., regarding the purposes of measuring a phenomenon, the emphasis is placed on its epistemic singularity. Accordingly, the normative power of a standard correlates with its specific performance or applicability in both measuring and validating research data.

The ideal of epistemic singularity of a standard could be literally taken from Witold Kula's phrase 'one king, one law, one weight, one measure'.[10] With this formula, Kula connects two major aspects of standard setting in his book on the history of measurement, namely the pre-existing authority ('the king', more recently, the 'International Bureau of Weights and Measures', BIPM, Paris) and the singularity of a standard gauge ('one weight', 'one measure').[11] As such, a standard format owns a specific, circumscribed and strongly regulated function. Hence, its use restricts us to a specific area of application(s). Standards are known for their binding character. There will be repercussions for ignoring as much as for dispensing standards. Under this directive, data that dispenses with current standards of research will be disregarded by the scientific community.

While addressing contemporary issues of standard setting, we have to review the existence or non-existence of an authority ('one king'). Especially when it comes to explicit fixation, the process of standard setting calls for such a pre-existing authority; or at least for an authority that is already acknowledged within a scientific community that calls for standard formats, for example the one concerning regulations of reporting scientific results. Given the ideal of explicit fixation, an authority is able to implement a new standard for scientific purposes. To aligning research practice, streamlining key processes of scientific activity could be regarded as major motives for implementing scientific norms. Explicit fixation is often accompanied by a veritable global perspective: here, setting and implementing standards comes with the objective to foster universal distribution and global application of formats that make it possible to provide uniform reference.

Against this background, current regimes of networking and distributing scientific authority are quite noteworthy: Especially, large research consortia of the 'Big Sciences' introduce new regimes of standard setting and implementing.[12] Here, standard formats, such as protocols of handling tissue, are the proverbial 'glue' that keeps working groups and individual researchers together. This development becomes apparent within the realm of research consortia, as for example, the Alzheimer's Disease Neuroimaging Initiative (ADNI) that launches and coordinates multicentre studies for research into Alzheimer's Disease.[13] Due to significant co-funding ADNI and other research consortia are often openly industry-oriented (in this case: the pharmaceutical industry). Besides having hidden agendas and motives for introducing a given set of research protocols, standards are key instruments of partnership creation and – maintenance: They do not only orientate multicentre studies, they also regulate, as the case of ADNI shows, the degree in which research data is distributed within the realm of a given network (database). This ratio impacts significantly on issues of participation: consortia often do not explicitly punish transgressions, as for example in the case of neglecting or misusing standard formats. Correctly adopting standard protocols at research sites is crucial for including local research data into the overall research design. Firstly, this could be regarded as a prerequisite for participating in a multicentre study. It may also pave the way for future cooperation. In order to address this issue properly, let us revisit the two types of implementing a standard format as a standard of research. In particular, it seems necessary to differentiate between further issues of endorsement. How is the epistemic rank of a standard maintained and affirmed? There are at least two answers to this question: In the first scenario a standard format is used continuously for a specific purpose (local practice). Accordingly, it is stabilized as scientific norm through application and frequent use. In the second scenario, a given standard format is strengthened by its degree of distribution (nonlocal practice). As the case of multicentre studies shows, scientific networks foster this process due to prerequisites for 'shared' protocols and other formats of standardization. Here, different aspects come together. In progress-stabilization and the distributed use of a standard format both reinforce the inherent authority of this highly scientific norm. Concurrently, the strengthened authority allows for further processes of endorsing and affirming a scientific norm as the true and only standard. Hence, we could address the relation of authority and endorsement as a self-vindicating circle.

Facing Challenges of Standardization

Historically, well-known challenges of standardization consist of a plurality of local weights and measures.[14] Consequently, introducing and enforcing standard units of weights and measures could be regarded as major goal of standardization. Within the realm of science, this is certainly the case for coming to grips with a standard experimental protocol, against the background of multiple experimental protocols that are debated upon, in order to inform about a certain phenomenon. Due to technical innovation and scientific progress, standards remain objects of improvement or even displacement. New standards and corresponding regulations are introduced if they have proved to be more potent with regard to the epistemic goals of a scientific endeavour, such as the measurement of temperature.[15] Additionally, different needs and local regimes of standardizing might bear a number of different standard formats. Each single format could be regarded as epistemic superior, while being compared with the other ones available. In the case of reference data (i.e. reference genome, reference brain) epistemic singularity is regularly aspired to, but not achieved. As a consequence, scientists are confronted with a plurality of standards. Taking the example of functional neuroimaging, different sets of reference data could be referred to, in order to tackle a given goal of measurement or comparison respectively.[16] In the following paragraphs, I shall address this issue under the heading of *conflicting standards*. The definition of conflicting standards is two or more standard formats of normalizing individual data exist (i.e., in functional neuroimaging). They are used interchangeably in a given field of research, because different 'laboratories' evaluate the efficacy of each of these standards differently, depending on research questions, the technical set-up (neuroimaging device), and further local preconditions of research. In the case of conflicting standards, scientists implement and further endorse parallel regimes of normalizing research data. With regard to visualizing data, the case is even worse: a study by Markus Christen and his colleagues, spanning fourteen years of display practice in functional neuroimaging, revealed a remarkably diverse use of colours (colour coding), and hence a plurality of ad-hoc standards which are used interchangeably.[17] With regard to data-driven approaches in the field of biomedicine, we may even have to acknowledge a multiplication of standards at every level of the research process, as Peter Keating and Alberto Cambrosio have pointed out.[18] Furthermore, databasing, to a significant degree, necessitates the development of inter-laboratory arrangements with regard to the technical set up and reporting practices. This is also the case in multicentre research trials, as for example with ADNI.[19]

Therefore, recent data-driven approaches in genetics, molecular biology, climate research and the physical sciences call for higher order standards. Firstly, higher order standards are norms that inform researchers about the standards (if there are more than one) that have to be consulted, and the way in which these are correctly adopted for a given experimental setting. In the case of research consortia, higher order standards are often built-in protocols that guide the implementation of a given experimental set-up and trace the handling of standard formats down to the most detailed steps. Take, for example, the choice of biological material (organisms), the inclusion/exclusion ratio of patients and further research subjects, technical equipment, parameters of testing, labelling system (i.e., nomenclature), ratio of data management and reporting practices. As inter-laboratory arrangements, they inform us about the major prerequisites that a local clinical research site has to fulfil in order to be a productive part of a multicentre study or the like. Secondly, higher order standards are developed for data sharing practices, ranging from protocols of data acquisition and data management, to standard formats of metadata for archiving and harvesting data for secondary or tertiary research purposes (cf. 'metadata standards').[20]

A specific challenge of contemporary research is this very need for defining higher order standards that inform and guide the choice of standard formats. For instance, metastandards for biological research data quality. such as the 'MIAME protocol', provide researchers with the *m*inimum *i*nformation *a*bout a *m*icroarray *e*xperiment, which is said to be 'needed to enable the interpretation of the results of the experiment unambiguously and potentially to reproduce the experiment'.[21] DNA microarray experiments are considered key tools in the post-genomic area, as one experiment involves the simultaneous analysis of many hundreds or thousands of genes.[22] The 2.0 version of the MIAME protocol defines six elements that ought to be provided to support microarray-based publications, which includes: (1) the raw data for each hybridization; (2) the final processed data for the set of hybridizations in the experiment (study); (3) the essential sample annotation, including experimental factors and their values; (4) the experimental design including sample data relationships; (5) sufficient annotation of the array design; and additionally (6) essential experimental and data processing protocols.[23]

As the case of MIAME illustrates, challenges of current approaches of standardizing research are deeply bound up with 'metastandardization'. In a way, the question of which standard format is pivotal for research practices is merely relegated to the realm of higher order standards. This may be for strictly pragmatic reasons, given that the exceptional quality of a singular standard is under debate. The question arises whether this practice impacts on the ratio of standard setting, or rather paves the way for a revised understanding of standardization itself. Firstly, key aspects of standardization are readdressed, namely: (a) on what grounds standards are prioritised (choice of standards); and (b) how they are implemented and who participates in this decision process, respectively (question of authority).[24] Secondly, metastandardization may implement a hierarchy of standards resulting in a hierarchy of standardization practices. Drawing upon Joan Fujimura's account of 'standardized packages',[25] it seems viable to elaborate on the question if we have to acknowledge a somewhat 'packaged', and in spite of the 'packaging', clustered organization of standardization. That is to say: we need to address the extent of how far metastandardization redefines each process of standard setting as a step in the very regime of standardization, all the way down to primary standards (e.g., reference data). It enables us to employ more detailed specifications of the software, supporting tools and more precise experimental descriptions (cf. MIAME). This immediately calls our attention to the question whether a given standard at one level (cluster) can be evaluated as separated from others at succeeding levels (clusters), or the other way around.

The very fact of generating and maintaining databases fosters the rise of inter-laboratory arrangements, which happens regularly. Several cases show that inter-laboratory arrangements might not be restricted to metastandardization, they may also include further regulatory mechanisms, for example, concerning the submission of raw data.[26] Finally, the case of higher order standards helps readdress and critically reflect on the regulatory power of standards per se. The classical model of standardization, which is oriented towards issues of stabilization and control (cf. calibration chain), is adopted in metastandardization. Hence, key challenges that stem from the classical model, as for example the dichotomic demand of providing uniform reference and allowing flexible adaption with regard to local prerequisites of scientific practices,[27] impact on regimes of metastandardization. Drawing upon metastandards for biological research data, it has been stated that standards must simultaneously appeal to very different fields of research, including those of industrial protagonists.[28] Unfortunately, 'these different groups do not necessarily share a consensus concerning the definition of what should count as "reliable" or "reproducible"'.[29]

Conclusion and Outlook

Drawing upon data-driven research in biomedicine and the life sciences, this paper has illustrated that the classical model of standardization, based on the ideal of the epistemic singularity of a scientific norm, is regularly aspired to, but not achieved. Accordingly, scientists are confronted with a plurality of scientific norms, local regimes, working standards and the like. One attempt at justifying this situation, on behalf of the classical model, would be that the rise of competing or conflicting standards should be regarded as an intermediate step of the very process of standardization itself. In contrast, an alternative attempt would account for the specific prerequisites of research in biomedicine and the life sciences – given the variability of biological entities and the plurality of research interests and methods respectively. Do either of these efforts comprehensively justify why scientists increasingly call for standardizing the choice and use of standards? Actually, the question of which standard format provides the key for a given research practice is often delegated to the realm of higher order standards. As regards the definition of higher order standards (metastandards), this may well be driven by pragmatic reasons in the first place, considering that the exceptional quality of a singular standard is under debate. Still, at the eve of an increasing prominence and influence of research consortia in biomedicine, it seems important to reflect on the very process of standardization of scientific practice itself. This not only includes the question, on what grounds standards are prioritized (choice of standards), but also how they are implemented, and who participates in this decision process, respectively (question of authority).

NOTES

Schlaudt and Huber, Introduction

1. This book emerged out of the conference 'Dimensions of Measurement' held at the Center for Interdisciplinary Studies (ZIF) in Bielefeld, Germany in March 2013 – as did its complementary volume in the same series *Reasoning in Measurement*, edited by Nicole Mößner and Alfred Nordmann.
2. I. Kant, *Metaphysical Foundations of Natural Science*, translated by M. Friedman, in *Theoretical Philosophy after 1781. The Cambridge Edition of the Works of Kant* (1786; Cambridge: University Press, 2002), p. 185.
3. E. Tal, 'Old and New Problems in Philosophy of Measurement', *Philosophy Compass*, 8:12 (2013), pp. 1159–73, on p. 1168.
4. L. Busch, *Standards. Recipes for Reality* (Cambridge, MA, and London: MIT Press, 2011).
5. S. Timmerman and M. Berg, *The Gold Standard. The Challenge of Evidence-Based Medicine and Standardization in Health Care* (Philadelphia, PA: Temple University Press, 2003).
6. W. E. K. Middleton, *A History of the Thermometer and its Uses in Meteorology* (Baltimore, MD: Johns Hopkins Press, 1966).
7. G. C. Bowker and S. L. Star, *Sorting Things Out. Classification and its Consequences* (Cambridge, MA, and London: MIT Press, 1999).
8. For example, M. Lampland and S. L. Star (eds), *Standards and their Stories. How Quantifying, Classifying, and Formalizing Practices Shape Everyday Life* (Ithaca, NY, and London: Cornell University Press, 2009).
9. J. H. Fujimura, 'Crafting Science: Standardized Packages, Boundary Objects, and Translation', in A. Pickering (ed.), *Science as Practice and Culture* (Chicago, IL, and London: University of Chicago Press, 1992), pp. 168–211.
10. A. Franklin, *Shifting Standards. Experiments in Particle Physics in the Twentieth Century* (Pittsburgh: University of Pittsburgh Press, 2013).
11. K. Mannheim, *Ideology and Utopia. An Introduction to the Sociology of Knowledge* (New York: Harcourt, Brace & Co; London: Routledge & Kegan Paul, 1954), p. 258.
12. C. Elgin (1989), 'The Relativity of Fact and the Objectivity of Value', in M. Krausz (ed.), *Relativism. Interpretation and Confrontation* (Notre Dame: University of Notre Dame Press, 1989), pp. 86–98, on p. 86. There is a more easily accessible reprint in *Harvard Review of Philosophy*, 6 (1996), pp. 4–15.
13. L. H. Nelson, *Who Knows: From Quine to a Feminist Empiricism* (Philadelphia, PA:

Temple University Press, 1990), H. Putnam, *The Collapse of the Fact/Value Dichotomy and Other Essays* (Cambridge, MA: Harvard University Press, 2002).

14. H. E. Longino, *The Fate of Knowledge* (Princeton, NJ: Princeton University Press, 2002).
15. T. J. Quinn and J. Kovalevsky, 'Measurement and Society', *Comptes Rendus Physique*, 5 (2004), pp. 791–7.

1 Kusch, 'A Branch of Human Natural History': Wittgenstein's Reflections on Metrology

1. B. van Fraassen, *Scientific Representation: Paradoxes of Perspective* (Oxford: Clarendon Press, 2008).
2. L. Wittgenstein, *Philosophical Grammar* (Oxford: Blackwell, 1974), p. 185.
3. F. Waismann, *The Principles of Linguistic Philosophy* (London: Macmillan, 1965), p. 14.
4. L. Wittgenstein, *Nachlass: The Bergen Electronic Edition* (Oxford: Oxford University Press, 2000), CD, ms. 109, p. 176.
5. G. E. Moore, 'Wittgenstein's Lectures in 1930–1933', in L. Wittgenstein, *Philosophical Occasions: 1912–1951* (Indianapolis, IN: Hackett, 1993), pp. 46–114, on p. 73.
6. L. Wittgenstein, *Wittgenstein's Lectures, Cambridge, 1932–1935* (Amherst, NY: Prometheus Books, 2001), p. 84.
7. L. Wittgenstein, *Remarks on the Foundations of Mathematics*, 3rd edn (Oxford: Blackwell, 1978), III: 36 (1939–40).
8. L. Wittgenstein, *Wittgenstein's Lectures on the Foundations of Mathematics, Cambridge 1939* (Chicago, IL: Chicago University Press, 1976), p. 104.
9. Wittgenstein, *Lectures on the Foundations of Mathematics*, pp. 113–14.
10. M. Kusch, 'Wittgenstein and Einstein's Clocks', in E. Ramharter (ed.), *Unsocial Sociabilities: Wittgenstein's Sources* (Berlin: Parerga, 2011), pp. 203–18.
11. M. Kusch, 'Kripke's Wittgenstein, *On Certainty*, and Epistemic Relativism', in D. Whiting (ed.), *The Later Wittgenstein on Language* (Basingstoke: Palgrave Macmillan, 2009), pp. 213–30.
12. L. Wittgenstein, *On Certainty* (Oxford: Blackwell, 1969), §492.
13. Wittgenstein, *On Certainty*, §305.
14. Wittgenstein, *Nachlass*, ms. 164, p. 82.
15. Wittgenstein, *Nachlass*, ms. 123, p. 19v.
16. Wittgenstein, *Nachlass*, ms. 113, p. 22v.
17. L. Wittgenstein, *Philosophical Investigations* (Oxford: Blackwell, 1953), §16.
18. L. Wittgenstein, *Nachlass: The Bergen Electronic Edition* (Oxford: Oxford University Press, 2000), CD, ms. 164, p. 82.
19. C. Penco, 'The Influence of Einstein on Wittgenstein's Philosophy', *Philosophical Investigations*, 33 (2010), pp. 360–79, on p. 375.
20. S. Kripke, *Wittgenstein on Rules and Private Language* (Cambridge, MA: Harvard University Press, 1982), p. 96.
21. L. Wittgenstein, *Philosophical Investigations* (Oxford: Blackwell, 1953), §206; L. Wittgenstein, 'Cause and Effect: Intuitive Awareness', in L. Wittgenstein, *Philosophical Occasions: 1912–1951* (Indianapolis, IN: Hackett, 1993), pp. 370–426.
22. Cf. Wittgenstein's Letter to Schlick from 20 November 1931: '…vielleicht der Hauptunterschied zwischen der Auffassung des Buches & meiner jetzigen ist, dass ich einsah,

dass die Analyse des Satzes nicht im Auffinden verborgener Dinge liegt, sondern im Tabulieren, in der übersichtlichen Darstellung, der Grammatik, d.h. des grammatischen Gebrauchs, der Wörter'. Noord-Hollands Archief, Haarlem, Wiener Kreis Archief, Moritz Schlick Nachlass, 123/Wittg-15.

23. L. Wittgenstein, *Wittgenstein's Lectures on the Foundations of Mathematics, Cambridge 1939* (Chicago, IL: Chicago University Press, 1976), pp. 104–5.

24. M. Kusch, *A Sceptical Guide to Meaning and Rules: Defending Kripke's Wittgenstein* (Chesham: Acumen, 2006).

25. Kripke, *Wittgenstein*.

26. D. Bloor, *Wittgenstein, Rules and Institutions* (London: Routledge, 1997).

27. Wittgenstein, *Nachlass*, ms. 222, p. 12.

28. Wittgenstein, *Lectures on the Foundations of Mathematics*, p. 105.

29. B. Barnes, D. Bloor and J. Henry, *Scientific Knowledge: A Sociological Analysis* (London: Athlone, 1996), pp. 55–9.

30. B. Barnes, D. Bloor and J. Henry, *Scientific Knowledge: A Sociological Analysis* (London: Athlone, 1996); B. Barnes and D. Bloor, 'Relativism, Rationalism and the Sociology of Knowledge', in M. Hollis and S. Lukes (eds), *Rationality and Relativism* (Oxford: Blackwell, 1982), pp. 21–47; P. Boghossian, *Fear of Knowledge: Against Relativism and Constructivism* (Oxford: Clarendon Press, 2006); G. Harman and J. Jarvis Thomson, *Moral Relativism and Moral Objectivity* (Oxford: Blackwell, 1996); G. Rosen, 'Nominalism, Naturalism, Epistemic Relativism', *Philosophical Perspectives*, 15 (2001), pp. 60–91; F. F. Schmitt, 'Introduction: Epistemic Relativism', *Episteme*, 4 (2007), pp. 1–9; B. Williams, 'The Truth in Relativism', in Williams, *Moral Luck* (Cambridge: Cambridge University Press, 1981), pp. 132–43; M. Williams, 'Why (Wittgensteinian) Contextualism Is Not Relativism', *Episteme*, 4 (2007), pp. 93–114.

31. Williams, 'Why (Wittgensteinian) Contextualism Is Not Relativism', p. 94.

32. L. Wittgenstein, *Remarks on the Foundations of Mathematics*, 3rd edn (Oxford: Blackwell, 1978), I: 150–8.

33. L. Wittgenstein, *Nachlass: The Bergen Electronic Edition* (Oxford: Oxford University Press, 2000), CD, ms. 142, p. 76.

34. L. Wittgenstein, *Denkbewegungen: Tagebücher 1930–1932, 1936–1937* (Frankfurt am Main: Fischer, 1999), p. 24.

35. Wittgenstein, *Nachlass*, ms. 159, p. 16v.

36. Wittgenstein, *Remarks on the Foundations of Mathematics*, I: 5.

37. L. Wittgenstein, *Zettel*, 2nd edn (Oxford: Blackwell, 1981), p. 388.

38. Williams, 'The Truth in Relativism'.

39. Barnes and Bloor, 'Relativism', p. 23.

40. Williams, 'The Truth in Relativism', pp. 141–2.

41. L. Wittgenstein, 'Remarks on Frazer's *Golden Bough*', in L. Wittgenstein, *Philosophical Occasions: 1912–1951* (Indianapolis, IN: Hackett, 1993), pp. 118–155.

42. L. Wittgenstein, *Denkbewegungen: Tagebücher 1930–1932, 1936–1937* (Frankfurt am Main: Fischer, 1999), p. 389.

43. L. Wittgenstein, *Wittgenstein's Lectures on the Foundations of Mathematics, Cambridge 1939* (Chicago, IL: Chicago University Press, 1976), p. 204.

44. L. Wittgenstein, *Wittgenstein's Lectures, Cambridge, 1932–1935* (Amherst, NY: Prometheus Books, 2001), p. 63.

45. L. Wittgenstein, *Philosophical Grammar* (Oxford: Blackwell, 1974), p. 322.

46. R. Rhees, 'Some Developments in Wittgenstein's View of Ethics', *Philosophical Review*,

74 (1965), pp. 17–26, on p. 23.

47. Rhees, 'Some Developments', p. 24.
48. Rhees, 'Some Developments', p. 22.
49. Wittgenstein, *Nachlass*, ms. 142, p. 75.
50. Wittgenstein, *Lectures on the Foundations of Mathematics*, 202.
51. Wittgenstein, *Denkbewegungen*, p. 75.
52. Wittgenstein, *Philosophical Investigations*, II, p. 192.
53. Wittgenstein, *Nachlass*, ms. 213, p. 236r.
54. Wittgenstein, *Remarks on the Foundations of Mathematics*, I: 118.
55. Wittgenstein, *Remarks on the Foundations of Mathematics*, VII: 18.
56. L. Wittgenstein, *Remarks on the Philosophy of Psychology*, 2 vols (Oxford: Blackwell, 1980), vol. 1, p. 47; vol. 2, pp. 658, 678–9; Wittgenstein, *Zettel*, pp. 365, 378–381.
57. B. van Fraassen, *Scientific Representation: Paradoxes of Perspective* (Oxford: Clarendon Press, 2008), p. 165.
58. Van Fraassen, *Scientific Representation*, p. 165.
59. Van Fraassen, *Scientific Representation*, p. 122.
60. Van Fraassen, *Scientific Representation*, p. 122.
61. Van Fraassen, *Scientific Representation*, p. 235.
62. Wittgenstein, *Remarks on the Foundations of Mathematics*, VII: 23.
63. L. Wittgenstein, *Philosophical Grammar* (Oxford: Blackwell, 1974), pp. 186–7.
64. S. Kripke, *Wittgenstein on Rules and Private Language* (Cambridge, MA: Harvard University Press, 1982).
65. L. Wittgenstein, *Nachlass: The Bergen Electronic Edition* (Oxford: Oxford University Press, 2000), CD, ms. 222, p. 12.
66. E. Durkheim, *The Elementary Forms of the Religious Life* (New York: The Free Press, 1965).

2 Schyfter, Metrology and Varieties of Making in Synthetic Biology

1. P. W. Bridgman, *The Logic of Modern Physics* (New York: Macmillan, 1932), p. 5.
2. J. L. Austin, *How to Do Things with Words* (Cambridge, MA: Harvard University Press, 1962).
3. M. Kusch, '"A Branch of Human Natural History": Wittgenstein's Reflections on Metrology', this volume.
4. Bridgman's work has received increased attention recently, and deserves further examination and use from fields such as science and technology studies and current history and philosophy of science. See for example: H. Chang, *Inventing Temperature: Measurement and Scientific Progress* (Oxford: Oxford University Press, 2004); H. Chang, 'Operationalism', *The Stanford Encyclopaedia of Philosophy* (Fall 2009), at http://plato.stanford.edu/archives/fall2009/entries/operationalism/ [accessed 31 March 2014].
5. A. Arkin, F. Arnold, D. Berry et al., 'Synthetic Biology: What's in a Name?', *Nature Biotechnology*, 27:12 (2009), pp. 1071–3.
6. M. A. O'Malley, A. Powell, J. F. Davies and J. Calvert, 'Knowledge-Making Distinctions in Synthetic Biology', *BioEssays*, 30:1 (2007), pp. 57–65.
7. Arkin et al., 'Synthetic Biology: What's in a Name?', p. 1071.
8. D. Endy, 'Foundations for Engineering Biology', *Nature*, 438:24 (2005), pp. 449–53, on p. 452. See also D. Endy, 'Synthetic Biology: Can We Make Biology Easy to Engineer?', *Industrial Biotechnology*, 4:4 (2008), pp. 340–51.

9. M. Heinemann and S. Panke, 'Synthetic Biology: Putting Engineering into Biology', *Bioinformatics*, 22:22 (2006), pp. 2790–9.

10. To maintain anonymity and ensure confidentiality, I use pseudonyms for all of my interviewees.

11. T. F. Gieryn, 'Boundary Work and the Demarcation of Science from Non-Science: Strains and Interests in Professional Ideologies of Scientists', *American Sociological Review*, 48:6 (1983), pp. 781–95.

12. M. Heinemann and S. Panke, 'Synthetic Biology: Putting Engineering into Biology', *Bioinformatics*, 22:22 (2006), pp. 2790–9, on p. 2790.

13. D. Endy, 'Foundations for Engineering Biology', *Nature*, 438:24 (2005), pp. 449–453, on p. 449.

14. See for example: D. Endy and A. Arkin, *A Standard Parts List for Biological Circuitry* (Berkeley, CA: Defense Advanced Research Projects Agency, 1999); L. H. Hartwell, J. J. Hopfield, S. Leibler and A. W. Murray, 'From Molecular to Modular Cell Biology', *Nature*, 402 (1999), pp. C47–C52; J. Lucks, L. Qi, W. R. Whitaker and A. Arkin, 'Towards Scalable Parts Families for Predictable Design of Biological Circuits', *Current Opinion in Microbiology*, 11:6 (2008), pp. 567–73; H. M. Sauro, 'Modularity Defined', *Molecular Systems Biology*, 4 (2008).

15. See for example: J. C. Anderson, J. E. Dueber, M. Leguia, G. C. Wu, J. A. Goler, A. P. Arkin and J. D. Keasling, 'BglBricks: A Flexible Standard for Biological Part Assembly', *Journal of Biological Engineering*, 4:1 (2010). A. Arkin, 'Setting the Standard in Synthetic Biology', *Nature Biotechnology*, 26:7 (2008), pp. 771–4; B. Canton, A. Labno and D. Endy, 'Refinement and Standardization of Synthetic Biological Parts and Devices', *Nature Biotechnology*, 26:7 (2008), pp. 787–93.

16. See for example: W. G. Vincenti, 'Control-Volume Analysis: A Difference in Thinking between Engineering and Physics', *Technology and Culture*, 23:2 (1982), pp. 145–74.

17. Heinemann and Panke, 'Synthetic Biology', p. 2791.

18. V. K. Mutalik, J. C. Guimaraes, G. Cambray et al., 'Quantitative Estimation of Activity and Quality for Collections of Functional Genetic Elements', *Nature Methods*, 10:4 (2013), pp. 347–53.

19. See W. G. Vincenti, 'The Air-Propeller Tests of W. F. Durand and E. P. Lesley: A Case Study in Technological Methodology', *Technology and Culture*, 20:4 (1979), pp. 712–51.

20. Vincenti, 'The Air-Propeller Tests of W. F. Durand and E. P. Lesley'.

21. M. Kusch, *Knowledge by Agreement* (Oxford: Oxford University Press, 2002), p. 2.

22. Kusch, *Knowledge by Agreement*, p. 162.

23. M. Kusch, '"A Branch of Human Natural History": Wittgenstein's Reflections on Metrology', in this volume.

24. M. Heinemann and S. Panke, 'Synthetic Biology: Putting Engineering into Biology', *Bioinformatics*, 22:22 (2006), pp. 2790–9, on p. 2792.

25. B. Canton, A. Labno and D. Endy, 'Refinement and Standardization of Synthetic Biological Parts and Devices', *Nature Biotechnology*, 26:7 (2008), pp. 787–93, on p. 787.

26. Kusch, *Knowledge by Agreement*, p. 163.

27. Kusch, *Knowledge by Agreement*, p. 164.

28. Kusch, *Knowledge by Agreement*, p. 164.

29. P. Schyfter, 'Technological Biology? Things and Kinds in Synthetic Biology', *Biology & Philosophy*, 27:1 (2012), pp. 29–48.

30. Schyfter, 'Technological Biology?'.

31. Heinemann and Panke, 'Synthetic Biology', p. 2792.
32. P. W. Bridgman, *The Logic of Modern Physics* (New York: Macmillan, 1932).

3 Plutniak, Refrain from Standards? French, Cavemen and Computers: A (Short) Story of Multidimensional Analysis in French Prehistoric Archaeology

1. F. Audouze and A. Leroi-Gourhan, 'France: A Continental Insularity', *World Archaeology*, 13:2 (1981), pp. 170–89.
2. L. R. Binford, 'Archaeology as Anthropology', *American antiquity*, 28:2 (1962), pp. 217–25.
3. J.-P. Benzécri, *L'analyse des données* (Paris: Dunod, 1976).
4. Namely: (Multiple) Correspondence Analysis, Multidimensional Scaling, Principal Component Analysis, etc. Benzecri among others: I. C. Lerman, M. Jambu...
5. L. R. Binford, 'A Preliminary Analysis of Functional Variability in the Mousterian of Levallois facies', *American Anthropology*, 68:2–2 (1966), pp. 238–95.
6. J. R. Goody, *The Domestication of the Savage Mind* (Cambridge: Cambridge University Press, 1977).
7. In the sense that the archaeological findings are materials and not verbals.
8. J.-C. Gardin, 'On a Possible Interpretation of Componential Analysis in Archeology', *American Anthropologist*, 67:5 (1965), pp. 9–22, on p. 20.
9. A. Cambrosio and P. Keating, 'The Disciplinary Stake: The Case of Chronobiology', *Social Studies of Science*, 13:3 (1983), pp. 323–53.
10. P. Keating, A. Cambrosio and M. MacKenzie, 'The Tools of the Discipline: Standards, Models, and Measures in the Affinity/Avidity Controversy in Immunology', in *The Right Tools for the Job. At Work in Twentieth-Century Life Sciences* (Princeton, NJ: Princeton University Press, 1992), pp. 312–54, 399.
11. Based on interviews, I assume that this journal can be considered as typical of the main trends of French prehistoric archaeology on this period. This field is faintly internationalized, as shown by the dominant majority of French-written papers.
12. French National Centre for Scientific Research.
13. G. Laplace, *Recherches sur l'origine et l'évolution des complexes leptolithiques* (Paris: E. de Boccard, 1966).
14. J.-É. Brochier and M. Livache, 'Le niveau C de l'abri no 1 de Chinchon à Saumanes de Vaucluse: analyse des correspondances et ses conséquences quant à l'origine des complexes du Tardiglaciaire en Vaucluse', *Géologie Méditerranéenne*, 5:4 (1978), pp. 359–69.
15. B. Bosselin and F. Djindjian, 'Une révision de la séquence de la Riera (Asturies) et la question du Badegoulien cantabrique', *Bulletin de la Société préhistorique française*, 96:2 (1997), pp. 153–73.
16. J. Dreyfus, *Implications ou neutralités des méthodes statistiques appliquées aux sciences humaines. L'analyse des correspondances* (Paris: CREDOC, 1975).
17. F. Bordes, 'Principes d'une méthode d'étude des techniques de débitage et de la typologie du Paléolithique ancien et moyen', *L'Anthropologie*, 54 (1950), pp. 19–34.
18. D. Sonneville-Bordes and J. Perrot, 'Essai d'adaptation des méthodes statistiques au Paléolithique supérieur. Premiers résultats', *Bulletin de la Société préhistorique de France*, 50:5–6 (1953), pp. 323–33.

19. Brochier and Livache, 'Le niveau C de l'abri no 1 de Chinchon à Saumanes de Vaucluse'.
20. My translation. A. Chollet, P. Boutin and B. Talur, 'Essai d'application des techniques de l'analyse des données aux pointes à dos des niveaux aziliens de Rochereil', *Bulletin de la Société préhistorique française*, 74:1 (1977), pp. 362–75, on p. 363.
21. Bosselin and Djindjian, Une révision de la séquence de la Riera'.
22. T. Shinn, 'New Sources of Radical Innovation: Research-Technologies, Transversality and Distributed Learning in a Post-Industrial Order', *Social Science Information*, 44:4 (2005), pp. 731–64, on p. 735.
23. D. R. Kelley, 'The Problem of Knowledge and the Concept of Discipline' in D. R. Kelley (ed.), *History and the Disciplines: The Reclassification of Knowledge in Early Modern Europe* (Rochester, NY: University of Rochester Press, 1997), pp. 13–28.
24. I shall introduce a difference between the syntactic/semantic orientation of the question raised, and the syntactic/semantic orientation of the method that is used for this purpose.
25. My translation. A. Chollet, P. Boutin and B. Talur, 'Essai d'application des techniques de l'analyse des données aux pointes à dos des niveaux aziliens de Rochereil', *Bulletin de la Société préhistorique française*, 74:1 (1977), pp. 362–75, on p. 375.
26. B. Bosselin and F. Djindjian, 'Un essai de structuration du Magdalénien français à partir de l'outillage lithique', *Bulletin de la Société préhistorique française*, 85:10–12 (1988), pp. 304–31.
27. G. Sauvet and S. Sauvet, 'Fonction sémiologique de l'art pariétal animalier franco-cantabrique', *Bulletin de la Société préhistorique française*, 76:10–12 (1979), pp. 340–54.
28. My translation. A. Decormeille and J. Hinout, 'Mise en évidence des différentes cultures Mésolithiques dans le Bassin Parisien par l'analyse des données', *Bulletin de la Société préhistorique française*, 79:3 (1982), pp. 81–8, on p. 83.
29. Decormeille and Hinout, 'Mise en évidence des différentes cultures Mésolithiques', p. 88.
30. Respectively, *'produits de première intention'* and *'produits d'intention dérivée'*. My translation. B. Bosselin, 'La séquence post-solutréenne du Parpalló (Espagne): application des méthodes quantitatives de l'analyse des données à l'étude morphométrique du débitage', *Bulletin de la Société préhistorique française*, 98:4 (2001), pp. 615–25, on p. 624.
31. A notion popularized by B. Latour, *Science in Action: How to Follow Scientists and Engineers through Society* (Cambridge, MA: Harvard University Press, 1987).
32. A. Marciniak, 'Setting a New Agenda. Ian Hodder and his Contribution to Archaeological Theory', *Archaeologia Polona*, 35–6 (1998), pp. 409–426, on p. 416.
33. In practice, this distinction is quite ambiguous; Gilles-Gaston Granger pointed out a 'formalist illusion', according to which the syntax could function without any relations with the semantic level: G.-G. Granger, *Pensée formelle et sciences de l'homme* (Paris: Aubier-Montaigne, 1967), p. 59.
34. A. Leroi-Gourhan, *Gesture and Speech*, trans. A. Bostock Berger (1st French edn: 1964; Cambridge, MA: MIT Press, 1993), p. 114.
35. A. Abbott, *Chaos of Disciplines* (Chicago, IL: University of Chicago Press, 2001), p. 59.
36. This is a long-standing method and its limits are well known. For an overview that dates back to the 1970s, see D. E. Chubin, 'State of the Field: The Conceptualization of Scientific Specialties', *Sociological Quarterly*, 17:4 (1976), pp. 448–76.
37. My sample is built from just one journal; however, I assume that this journal is central enough in this field to give access to a significant part of the literature.

38. Google Scholar's indexation appears as the best approximation, especially for this kind of French-written literature characterized by a low citation and diffusion rate.
39. For a dynamic modelization of the development of a scientific specialty, where education and publications have a key function, see: N. C. Mullins, 'The Development of a Scientific Specialty: The Phage Group and the Origins of Molecular Biology', *Minerva*, 10:1 (1972), pp. 51–82.
40. A. Abbott, *Chaos of Disciplines* (Chicago, IL: University of Chicago Press, 2001).
41. Contrary to the English-speaking custom, archaeometry in France includes mainly physical and chemical analysis of archaeological remains and far less data computation.
42. F. Djindjian, 'The Golden Years for Mathematics and Computers in Archaeology (1965–1985)', *Archeologia e Calcolatori*, 20 (2009), pp. 61–73.
43. M. J. Baxter, 'Mathematics, Statistics and Archaeometry: The Past 50 Years or So', *Archaeometry*, 50:6 (2008), pp. 968–82.
44. J.-É. Brochier, 'Plus c'est long, plus c'est large … mais encore? Sur quelques caractères métriques des lames de plein débitage', *Archéologies de Provence et d'ailleurs. Mélanges offerts à Gaëtan Congès et Gérard Sauzade.* (Aix-en-Provence: Bulletin archéologique de Provence, Supplément 5, Éditions de l'APA, 2008), pp. 75–86.
45. W. Stoczkowski, *Explaining Human Origins: Myth, Imagination and Conjecture* (1st French edn: 1994; Cambridge: Cambridge University Press, 2002).

4 de Courtenay, The Double Interpretation of the Equations of Physics and the Quest for Common Meanings

1. J. D'Alembert, *Traité de dynamique*, cited by J. Roche, *The Mathematics of Measurement. A Critical History* (London: The Athlone Press, 1998), p. 126.
2. J. C. Maxwell, 'Dimensions', in T. S. Baynes (ed.) *Encyclopedia Britannica*, 9th edn, 25 vols (Edinburgh: A. and C. Black, 1875-1889), vol. 7, pp. 240–1, on p. 241.
3. For a thorough exposition of this change, see J. Roche, *The Mathematics of Measurement. A Critical History* (London: The Athlone Press, 1998), pp. 145–62.
4. I. Hacking, 'The Granary of Science', in *The Taming of Chance* (Cambridge: Cambridge University Press, 1990), pp. 55–63.
5. Roche, *The Mathematics of Measurement*, p. 150.
6. P. Duhem, *La Théorie physique. Son objet, sa structure*, 2nd edn (Paris: Vrin, 1981), p. 199; also R. Carnap, *Physikalische Begriffsbildung* (Karlsruhe: Braun, 1926).
7. This account does not cover the later period of logical positivism in which the conception of testability, and therefore of empirical meaning, was modified.
8. Descombes enlarges on Taylor's ideas in V. Descombes, *Les institutions du sens* (Paris: Les Éditions de minuit), pp. 291–308.
9. H. Longino, *The Fate of Knowledge* (Princeton, NJ: Princeton University Press, 2002).
10. J. Fourier, *Théorie analytique de la chaleur* (Paris: Firmin Didot, 1822), § 161.
11. It should be noted that Fourier's investigation is limited because he only considers changes involving the size of the units.
12. They are 'secondary' quantities because they are quantities derived from primary quantities one already knows how to measure independently. We will see below that it would be more correct to use, in this section, the term 'magnitude' instead of 'quantity'.
13. J. C. Maxwell, 'Remarks on the Mathematical Classification of Physical Quantities', *Proceedings of the London Mathematical Society* (1871) s1–3(1), pp. 224–33, on p. 225.

14. On the importance of symbolic algebra in England, C. Smith and M. N. Wise, *Energy and Empire. A Biographical Study of Lord Kelvin* (Cambridge: Cambridge University Press, 1989), pp. 149–202.

15. A. Lodge, 'The Multiplication and Division of Concrete Quantities', *Nature*, 38 (1888), pp. 281–83, on p. 281.

16. See C. Smith and M. N. Wise, *Energy and Empire*, pp. 168–92.

17. On the difference between the Cartesian and the vectorial style in mathematics, see G. G. Granger, *Essai d'une philosophie du style* (Paris: Odile Jacob, 1988), pp. 71–105.

18. Maxwell, 'On the Mathematical Classification of Physical Quantities', p. 226.

19. J. De Boer, 'On the History of the Quantity Calculus and the International System', *Metrologia*, 31 (1994/95), pp. 405–29.

20. F. B. Silsbee, 'Systems of Electrical Units', *Journal of Research of the National Bureau of Standards-C. Engineering and Instrumentation*, 66C:2 (1962), pp. 137–83, on p. 141.

21. Silsbee, 'System of Electrical Units', p. 141.

22. Silsbee, 'System of Electrical Units', p. 153; De Boer, 'On the History of the Quantity Calculus and the International System', p. 418.

23. C. H. Page, 'The Mathematical Representation of Physical Entities', *IEEE Transactions on Education*, E-10:2 (1967), pp. 70–4, on p. 71.

24. H. König, 'Ueber die Mehrdeutigkeit des Grössenbegriffs', *Bulletin de l'Association suisse des électriciens*, 41 (1950), pp. 625–9; Silsbee, 'System of Electrical Units', p. 141.

25. B. Russell, *The Principles of Mathematics* (New York: W. W. Norton & Company, Inc. Publishers, 1938), p. 159.

26. See De Boer, 'On the History of the Quantity Calculus and the International System', pp. 421–4.

27. I echo here R. Saint-Guilhem's explanations that show how the magnitude equations can be reconstructed out of measure equations. R. Saint-Guilhem, 'Système d'unités et analyse dimensionnelle', *Annales des Mines*, 137 (1948), pp. 9–41.

28. R. Nozic, 'Invariance and Objectivity', in R. Nozic, *Invariances. The Structure of the Objective World* (Cambridge Mass.: The Belknap Press of Havard University Press, 2001), pp. 75–119, on p. 80.

29. Nozic, 'Invariance and Objectivity', p. 79.

30. See J. C. C. McKinsey and P. Suppes, 'On the Notion of Invariance in Classical Mechanics', *The British Journal for the Philosophy of Science*, 5:20 (1955), pp. 290–302, for their discussion of Galilean invariance in classical physics.

31. R. Giere, 'Using Models to Represent Reality', in L. Magnani, N. J. Nersessian and P. Thagard (eds), *Model-Based Reasoning in Scientific Discovery* (New York: Kluwer Academic Plenum Publishers, 1999), pp. 41–57.

5 Mari, An Overview of the Current Status of Measurement Science: From the Standpoint of the *International Vocabulary of Metrology* (*VIM*)

1. The author is a member of the Joint Committee on Guides in Metrology (JCGM), Working Group 2 (*VIM*). The opinion expressed in this paper does not necessarily represent the view of this Working Group.

2. The source of measurability in a formal concept of quantity is the explicit assumption of the seminal paper O. Hölder, 'Die Axiome der Quantität und die Lehre vom

Mass' (1901), translated from the German by J. Michell and C. Ernst as 'The Axioms of Quantity and the Theory of Measurement', *Journal of Mathematical Psychology*, 40 (1996), pp. 235–52. In fact, until a relatively recent past measurability was considered a feature specific of geometric quantities, so that expressions such as 'weights and measures' (as in 'International Bureau of Weights and Measures') or 'counting and measuring' (as in the paper by H. Helmholtz, 'Zählen und Messen erkenntnis-theoretisch betrachtet' (1887), translated from the German by C. L. Bryan as *Counting and Measuring* (Princeton, NJ: Van Nostrand, 1930)) are explained under the assumption that weights are not 'measures' and counting is not 'measuring'. An interesting, critical analysis on this position can be found in M. Bunge, 'On Confusing "Measure" with "Measurement" in the Methodology of Behavioral Science', in M. Bunge (ed.), *The Methodological Unity of Science* (Dordrecht: Reidel, 1973), pp. 105–22.

3. For example, the reference book F. S. Roberts, *Measurement Theory – With Applications to Decision-Making, Utility and Social Sciences* (London: Addison-Wesley, 1979) has been published in the Encyclopedia of Mathematics and its Applications series and no concessions to experimental topics have been given in L. Narens, *Abstract Measurement Theory* (Cambridge, MA: MIT Press, 1985).

4. A synthetic conceptual reconstruction of the options around the definition of 'measurement' is in L. Mari, 'A Quest for the Definition of Measurement', *Measurement*, 46 (2013), pp. 2889–95.

5. From the introduction of ISO, *International Vocabulary of Basic and General Terms in Metrology (VIM)*, International Bureau of Weights and Measures (BIPM), International Electrotechnical Commission (IEC), International Organization for Standardization (ISO), International Organization of Legal Metrology (OIML) (Geneva: ISO, 1984).

6. ISO, *VIM*, introduction.

7. P. M. Clifford, 'The International Vocabulary of Basic and General Terms in metrology', *Measurement*, 3 (1985), pp. 72–6.

8. ISO, *VIM*, introduction.

9. ISO/IEC Guide 2:2004, *Standardization and Related Activities – General Vocabulary*, 8th edn (Geneva: ISO/IEC, 2004). A note to the quoted definition explicitly states: 'Consensus need not imply unanimity'.

10. ISO/IEC DIR 1, *ISO/IEC Directives – Part 1: Procedures for the Technical Work*, 10th edn (Geneva: ISO/IEC, 2013).

11. The concepts of terminological dictionary and vocabulary are defined in ISO 1087–1:2000, *Terminology Work – Vocabulary – Part 1: Theory and Application* (Geneva: ISO, 2000).

12. JCGM, *VIM*, Scope.

13. JCGM, *VIM*, Conventions – Terminology rules.

14. ISO, *VIM*.

15. ISO, *VIM*, International Bureau of Weights and Measures (BIPM) and other six international organizations, 2nd edition (Geneva: ISO, 1993).

16. JCGM 200:2012, *International Vocabulary of Metrology – Basic and General Concepts and Associated Terms (VIM)*, 3rd edn (2008 version with minor corrections; Sèvres: Joint Committee for Guides in Metrology, 2012), at http://www.bipm.org/en/publications/guides/vim.html [accessed 7 April 2015]. The current, third edition of the *VIM* was initially published in 2007 as ISO/IEC Guide 99. In 2008 it was formally identified as a JCGM document and made freely downloadable from the web site of the BIPM. All the unreferenced quotations in the present paper are taken from the *VIM3*.

17. JCGM, *VIM*, introduction.

18. JCGM, *Charter – Joint Committee for Guides in Metrology* (2009), at http://www.bipm.org/utils/en/pdf/JCGM-charter.pdf [accessed 7 April 2015].

19. The nature of quantity values is controversial, as they are considered sometimes empirical entities sometimes linguistic ones. See our analysis in L. Mari and A. Giordani, 'Quantity and Quantity Value', *Metrologia*, 49 (2012), pp. 756–64.

20. S. S. Stevens, 'On the Theory of Scales of Measurement', *Science*, 103:2684 (1946), pp. 677–80.

21. A. Giordani and L. Mari, 'Property Evaluation Types', *Measurement*, 45 (2012), pp. 437–52.

22. T. S. Kuhn, *The Structure of Scientific Revolutions* (Chicago, IL: University of Chicago Press, 1970).

23. N. Hanson, *Patterns of Discovery* (Cambridge: Cambridge University Press, 1958).

24. N. Goodman, *Ways of Worldmaking* (Indianapolis, IN: Hackett, 1978).

25. JCGM 100:2008, *Evaluation of Measurement Data – Guide to the Expression of Uncertainty in Measurement* (*GUM*) (1993; Sèvres: JCGM, 2008), D.3.4, at http://www.bipm.org/en/publications/guides/gum.html [accessed 7 April 2015].

26. H. Margenau, 'Philosophical Problems Concerning the Meaning of Measurement in Physics', *Philosophy of Science*, 25:1 (1958), pp. 23–33.

27. The concept of model is a complex and elusive one. What is defined by the *VIM3*, that would have been better termed 'mathematical model of transduction', well complies with the simple and synthetic characterization: 'To an observer B, an object A* is a model of an object A to the extent that B can use A* to answer questions that interest him about A. The model relation is inherently ternary. Any attempt to suppress the role of the intentions of the investigator B leads to circular definitions or to ambiguities about "essential features" and the like', from M. Minsky, 'Matter, Mind and Models', in *Semantic Information Processing* (Cambridge: MIT Press, 1968), pp. 425–32, on p. 425.

28. The *VIM3* definition is actually 'quantity value provided...', not 'quantity provided...'. This appears to be just a mistake.

29. The concept of observability is very delicate for its long history of diverse usages in the philosophy of science. We can operationalize it, and avoid any condition of anthropomorphism, by simply assuming that a quantity is observable if the process of attributing a value to it is a *primitively solvable* problem, as for direct comparison of quantities, or counting of easily identifiable entities, or recognition of set membership (the latter case justifies the widespread adoption of analogue-to-digital converters in series to sensors, so to transform the indication to a selector among pre-identified quantized channels).

30. JCGM, *GUM*, 0.1.

31. L. Mari, P. Carbone and D. Petri, 'Measurement Fundamentals: A Pragmatic view', *IEEE Trans. Instr. Meas.*, 61:8 (2012), pp. 2107–15.

6 Grégis, Can We Dispense with the Notion of 'True Value' in Metrology?

1. JCGM, *Evaluation of Measurement Data – Guide to the Expression of Uncertainty in Measurement* (*GUM*) (1993; Sèvres: JCGM, 2008), at http://www.bipm.org/en/publications/guides/gum.html [accessed 7 April 2015].

2. JCGM, *International Vocabulary of Metrology – Basic and General Concepts and Associated Terms (VIM)*, 3rd edn (2008 version with minor corrections; Sèvres: Joint Committee for Guides in Metrology, 2012), at http://www.bipm.org/en/publications/guides/vim.html [accessed 7 April 2015].

3. See for example the *VIM* for a current definition of the term: JCGM, *International Vocabulary of Metrology – Basic and General Concepts and Associated Terms (VIM)*, 3rd edn (2008 version with minor corrections; Sèvres: Joint Committee for Guides in Metrology, 2012), at http://www.bipm.org/en/publications/guides/vim.html [accessed 7 April 2015], p. 22.

4. JCGM, *VIM*, p. viii.

5. JCGM, *Evaluation of Measurement Data – Guide to the Expression of Uncertainty in Measurement (GUM)* (1993; Sèvres: JCGM, 2008), at http://www.bipm.org/en/publications/guides/gum.html [accessed 7 April 2015], p. 3.

6. I refer here to Tal's five notions of measurement accuracy. The meaning of measurement accuracy in metrology corresponds to Tal's 'metaphysical accuracy': E. Tal, 'How Accurate Is the Standard Second?', *Philosophy of Science*, 78 (2011), pp. 1082–96.

7. JCGM, *VIM*, p. x. At this point we can identify 'measurand' with 'quantity'. We will see later how the two terms differ. See L. Mari's contribution to this volume for a more detailed development on the notion of 'measurand'.

8. ISO, *International Vocabulary of Basic and General Terms in Metrology (VIM)* (Geneva: ISO, 1984), p. 16.

9. JCGM, *VIM*, p. 25.

10. This move was already tackled in L. Mari, 'Epistemology of Measurement', *Measurement*, 34 (2003), pp. 17–30, especially on p. 18.

11. S. Stigler, *The History of Statistics: The Measurement of Uncertainty before 1900* (Cambridge, MA: Belknap Press of Harvard University Press, 1986), p. 158.

12. W. Bich, 'From Errors to Probability Density Functions. Evolution of the Concept of Measurement Uncertainty', *IEEE Transactions on Instrumentation and Measurement*, 61 (2012), pp. 2153–9, on pp. 2155–6: '[the classical] attitude is based on the illusion that subjectivity can be totally avoided in measurement, whereas it permeates much of it'.

13. See for example W. T. Estler, 'Measurement as Inference: Fundamental Ideas', *Annals of the CIRP*, 48 (1999), pp. 611–31, on p. 618: 'A great deal of time can be wasted in heated arguments concerning the exact form of the [probability density], which describes not reality in itself but only one's knowledge about reality'.

14. For a recent attempt to defend the traditional approach by resolving its internal difficulties, see R. Willink, *Measurement Uncertainty and Probability* (Cambridge: Cambridge University Press, 2013), especially pp. 72–81.

15. C. Ehrlich, R. Dybkaer and W. Wöger, 'Evolution of Philosophy and Description of Measurement (Preliminary Rationale for *VIM3*)', *Accreditation and Quality Assurance*, 12 (2007), pp. 201–18, on p. 209.

16. JCGM, *VIM*, p. 16.

17. I. Lira, *Evaluating the Measurement Uncertainty: Fundamental and Practical Guidance* (Bristol and Philadelphia, PA: Institute of Physics Publishing, 2002), p. 176.

18. Some alternative approaches try to make the true value disappear in the equations themselves. See for example the *IEC* approach described in Ehrlich et al., 'Evolution of Philosophy and Description of Measurement', pp. 213–17.

19. C. Eisenhart, 'Realistic Evaluation of the Precision and Accuracy of Instruments Calibration Systems', *Journal of Research of the National Bureau of Standards – C.*

Engineering and Instrumentation, 67C (1963), pp. 161–87, on p. 171. See also Willink, *Measurement Uncertainty and Probability*, p. 6.

20. C. Robert, and J. Treiner, 'Incertitudes des mesures de grandeurs', in J.-P. Kahane (ed), *Commission de réflexion sur l'enseignement des mathématiques, annexe sur la statistique* (2003), pp. 6–17, on p. 8.

21. 'The term "true value of a measurand" … is avoided in this Guide because the word "true" is viewed as redundant', JCGM, *GUM*, p. 50. The unknown true value of a quantity contrasts with the actually assigned value of that quantity, the latter being the known end product of a measurement process, typically destined to theory testing or decision making. In the *GUM*, any reference to a true value is dismissed (for the reasons already mentioned in this paper), and the emphasis is put on the actually assigned value (see p. 59). Since no differentiation between the true value and the assigned value is then needed any more, the choice was made in the *GUM* to designate the assigned value directly by the mere term 'value'. However, this leads to great misunderstandings and to a lack of clarity since, as we showed, an equivalent to the true value is needed, at least as a model parameter, in the value attribution processes. The choice made in the 2008 edition of the VIM to maintain the traditional categories '(quantity) value' and 'true (quantity) value' seems, in this regard, a much better decision.

22. W. C. Wimsatt, Re-Engineering Philosophy for Limited Beings: Piecewise Approximations to Reality (Cambridge, MA: Harvard University Press, 2007), pp. 94–5.

23. This is the definition of the 1993 edition of the *VIM*, see ISO, *International Vocabulary of Basic and General Terms in Metrology* (*VIM*) (Geneva: ISO, 1984), p. 20.

24. JCGM, *International Vocabulary of Metrology – Basic and General Concepts and Associated Terms* (*VIM*), 3rd edn (2008 version with minor corrections; Sèvres: Joint Committee for Guides in Metrology, 2012), at http://www.bipm.org/en/publications/ guides/vim.html [accessed 7 April 2015], p. 17.

25. JCGM, *Evaluation of Measurement Data – Guide to the Expression of Uncertainty in Measurement* (*GUM*) (1993; Sèvres: JCGM, 2008), at http://www.bipm.org/en/publications/guides/gum.html [accessed 7 April 2015], pp. 49–50.

26. JCGM, *VIM*, p. 20.

27. JCGM, *VIM*, p. 25.

28. See for example C. Eisenhart, 'Realistic Evaluation of the Precision and Accuracy of Instruments Calibration Systems', *Journal of Research of the National Bureau of Standards – C. Engineering and Instrumentation*, 67C (1963), pp. 161–87, on p. 171.

29. JCGM, *GUM*, p. 1.

30. As underlines Treiner, 'uncertainties related to the variability of phenomena … are stable and do not decrease when our knowledge progresses', in J. Treiner, 'Variabilité, incertitude, erreur', *Bulletin d'Union des Physiciens*, 105 (2011), pp. 9–14, on p. 14 (my translation).

31. ISO, *VIM*, p. 16.

32. See also Mari's objections in L. Mari, 'Epistemology of Measurement', *Measurement*, 34 (2003), pp. 17–30, on p. 22.

33. See N. De Courtenay's contribution to this volume for a development on the dual role of physical equations.

34. A. Giordani and L. Mari, 'Measurement, Models, and Uncertainty', *IEEE Transactions on Instrumentation and Measurement*, 61 (2012), pp. 2144–52, on p. 2147.

35. This reflects the conception of 'mathematical idealisation' described in E. McMullin, 'Galilean Idealization', *Studies in History and Philosophy of Science*, 16 (1985), pp. 247–73, on pp. 248–54.

36. I owe this example to Marc Priel, from the LNE (Laboratoire National de métrologie et d'Essais), Paris.

37. It should be mentioned here that Tal has identified quantity individuation among three major epistemological problems about measurement: E. Tal, 'The Epistemology of Measurement: A Model-Based Account' (PhD dissertation, University of Toronto, 2012), pp. 48–92. This is not the issue that I address here, although I hold definitional uncertainty, and the concept of 'measurand', to be strongly relevant to it.

38. JCGM, *VIM*, p. 20.

39. See for example Giordani and Mari, 'Measurement, Models, and Uncertainty', pp. 2147–9.

40. Giordani and Mari, 'Measurement, Models, and Uncertainty', p. 2148.

41. E. Tal, 'How Accurate Is the Standard Second?', *Philosophy of Science*, 78 (2011), pp. 1082–96, on p. 1090.

42. JCGM, *GUM*, p. 4.

43. As underlines Treiner, 'physical quantities only have a meaning at a certain scale and for a given use', in J. Treiner, 'Variabilité, incertitude, erreur', *Bulletin d'Union des Physiciens*, 105 (2011), pp. 9–14, on p. 12 (my translation).

44. M. Priel, 'Guide du vocabulaire de la métrologie: le concept de la valeur vraie fait débat', *Mesures*, 804 (2008), pp. 20–3, on p. 20.

45. J. Michell, 'The Logic of Measurement: A Realist Overview', *Measurement*, 38 (2005), pp. 285–94.

46. See L. Mari, 'The Role of Determination and Assignment in Measurement', *Measurement*, 21 (1997), pp. 79–90 as well as L. Mari, 'Beyond the Representational Viewpoint: A New Formalization of Measurement', *Measurement*, 27 (2000), pp. 71–84 and L. Mari, 'The Problem of Foundations of Measurement', *Measurement*, 38 (2005), pp. 259–66.

7 Soler, Calibration in Scientific Practices which Explore Poorly Understood Phenomena or Invent New Instruments

1. PratiScienS stands for 'Rethinking Science from the Standpoint of Scientific Practices'. See http://poincare.univ-lorraine.fr/fr/operations/pratisciens/accueil-pratisciens [accessed 27 March 2015]. The PratiScienS project has been supported by the ANR (Agence Nationale de la Recherche), MSHL (Maison des Sciences de l'Homme Lorraine), Région Lorraine and LHSP – Laboratoire d'Histoire des Sciences et de Philosophie – Archives Henri Poincaré (UMR 7117 of the CNRS).

2. L. Soler, F. Wieber, C. Allamel-Raffin, J. Gangloff, C. Dufour and E. Trizio (2013), 'Calibration: A Conceptual Framework Applied to Scientific Practices Which Investigate Natural Phenomena by Means of Standardized Instruments', *Journal for the General Philosophy of Science*, 44 (2013), pp. 263–317. See this paper for further details on the conceptual framework which is here only briefly summarized, as well as for bibliographic elements on calibration in the science studies.

3. The present paper benefitted from discussion with audiences at the International Conference 'Dimensions of Measurement' organized by A. Nordmann and O. Schlaudt in Bielefeld (March 2013).

4. Roughly presented in L. Soler, F. Wieber, C. Allamel-Raffin, J. Gangloff, C. Dufour and E. Trizio (2013), 'Calibration: A Conceptual Framework Applied to Scientific Prac-

tices Which Investigate Natural Phenomena by Means of Standardized Instruments', *Journal for the General Philosophy of Science*, 44 (2013), pp. 263–317.

5. See Soler et al, 'Calibration'.
6. H. M. Collins, *Changing Order: Replication and Induction in Scientific Practice* (1985; Chicago, IL: University of Chicago Press, 1992), pp. 100–6.
7. Collins, *Changing Order*, pp. 104–5.
8. For illustrations, see the references given in L. Soler, F. Wieber, C. Allamel-Raffin, J. Gangloff, C. Dufour and E. Trizio (2013), 'Calibration: A Conceptual Framework Applied to Scientific Practices Which Investigate Natural Phenomena by Means of Standardized Instruments', *Journal for the General Philosophy of Science*, 44 (2013), pp. 263–317, sections 2.3.2 and 2.2.4.
9. See, e.g., J. Woodward, 'Data and Phenomena', *Synthese*, 79 (1989), pp. 393–472, or N. Rasmussen, 'Facts, Artefacts, and Mesosomes: Practicing Epistemology with the Electron Microscope', *Studies in History and Philosophy of Science* (Part A), 24:2 (1993), pp. 227–65.
10. See, e.g., Rasmussen 'Facts, Artefacts, and Mesosomes', who explicitly categorizes some episodes involving the two models of microscopes under the heading of 'calibration'.

8 Katzir, Time Standards from Acoustic to Radio: The First Electronic Clock

1. I wrote this article while I enjoyed a Marie Curie fellowship of the Gerda Henkel Foundation.
2. S. Schaffer, 'Rayleigh and the Establishment of Electrical Standards', *European Journal of Physics*, 15 (1994), pp. 277–85; B. J. Hunt, 'The Ohm Is Where the Art Is: British Telegraph Engineers and the Development of Electrical Standards', *Osiris*, 9 (1994), pp. 48–63; A. Hessenbruch, 'Calibration and Work in the X-Ray Economy, 1896–1928', *Social Studies of Science*, 30 (2000), pp. 397–420.
3. On the ethos of exactitude see M. N. Wise (ed.), *The Values of Precision* (Princeton, NJ: Princeton University Press, 1995); J. L. Heilbron, *Weighing Imponderables and Other Quantitative Science around 1800* (Berkeley, CA: University of California Press, 1993).
4. M. W. Jackson, *Harmonious Triads: Physicists, Musicians, and Instrument Makers in Nineteenth-Century Germany* (Cambridge, MA: MIT, 2008).
5. L. S. Reich, *The Making of American Industrial Research: Science and Business at GE and Bell, 1876–1926* (Cambridge: Cambridge University Press, 1985); M. Amoudry, *Le général Ferrié et la naissance des transmissions et de la radiodiffusion* (Grenoble: Presses universitaires Grenoble, 1993), pp. 162–95.
6. Anon., *Circular of the Bureau of Standards, No. 74: Radio Instruments and Measurements* (Washington: Government Printing Office, 1918), pp. 96–109; H. Abraham and E. Bloch, 'Amplificateurs pour courants continus et pour courants de très basse fréquence', *Académie des sciences (France). Comptes rendus* (hereafter *Comptes rendus*), 168 (1919), pp. 1105–8, on p. 1106.
7. H. Abraham and E. Bloch, 'Mesure en valeur absolue des périodes des oscillations électriques de haute fréquence', *Journal de physique théorique et appliquée*, 9 (1919), pp. 211–22, on pp. 212–13. They described their results in confidential reports in 1916 and 1917.
8. D. Pantalony, *Altered Sensations: Rudolph Koenig's Acoustical Workshop in Nineteenth-*

century Paris (Dordrecht: Springer, 2009), pp. 22–25, and passim; J. H. Ku, 'Uses and Forms of Instruments: Resonator and Tuning Fork in Rayleigh's Acoustical Experiments', *Annals of Science*, 66 (2009), pp. 371–95.

9.   Abraham and Bloch, 'Mesure en valeur absolue'.

10. H. Abraham and E. Bloch, 'Entretien des oscillations méchaniques au moyen des lampes à trois électrodes', *Comptes rendus*, 168 (1919), pp. 1197–8; H. Abraham and E. Bloch, 'Entretien des oscillations d'un pendule ou d'un diapason avec un amplificateur à lampes', *Journal de physique théorique et appliquée*, 9 (1919), pp. 225–33.

11. Normally, triode valve circuits oscillated at much higher frequencies.

12. On electric clocks see, e.g., M. Viredaz, 'Horloges électriques', at http://www. chronometrophilia.ch/Electric-clocks/francais.htm [accessed 28 February 2014]; R. Stevenson, 'Mechanical and Electrical Clocks', in D. Howse (ed.), *Greenwich Time and the Discovery of the Longitude* (Oxford: Oxford University Press, 1980), pp. 213–19. Ku, 'Uses and Form', p. 376.

13. M. Lavet, 'Propriétés des organes électromagnétiques convenant aux petits moteurs chronométriques a diapason', *Annales Françaises de chronométrie*, 15 (1961), pp. 183–96; N. Hulin, 'Un mutualisme pédagogique au tournant des XIXe et XXe siècles. Informer, échanger, centraliser', at http://www.aseiste.org/documents/53f9941495b eb59b3ecf87631053f420.pdf [accessed 26 June 2012]; A. and V. Guillet, 'Nouveaux modes d'entretien des diapasons', *Comptes rendus*, 130 (1900), pp. 1002–4; A. Guillet, 'Roue à denture harmonique, application à la construction d'un chronomètre de laboratoire à mouvement uniforme et continu', *Comptes rendus*, 160 (1915), pp. 235–7; see also the list of Guillet's publications in Web of Science. On Lippmann's metrology, see D. J. Mitchell, 'Measurement in French Experimental Physics from Regnault to Lippmann: Rhetoric and Theoretical Practice', *Annals of Science*, 69 (2012), pp. 453–82.

14. J. A. Ratcliffe, 'William Henry Eccles. 1875–1966', *Biographical Memoirs of Fellows of the Royal Society*, 17 (1971), pp. 195–214, quotations on p. 198; W. H. Eccles, 'The Use of the Triode Valve in Maintaining the Vibration of a Tuning Fork', *Proceedings of the Physical Society of London*, 31 (1919), p. 269

15. Jordan's affiliation is mentioned in his patents, e.g., 'Improvements in application of thermionic valves to production of alternating currents and in relaying'. GB 155854, filed 17 April 1918. 'A list of the principal reports of experiment and investigation received by the Board of Invention and Research from August 1915 to February 1918', National Archives Britain (hereafter NAB), ADM 293/21, also in 212/159.

16. A. Korn, *Elektrische Fernphotographie und Ähnliches*, 2nd edn (Leipzig: S. Hirzel, 1907), pp. 66–9. Eccles's recollection quoted in Ractliffe 'Eccles', p. 198; 'Fourth Periodical Schedule of Reports Concerning Experiment and Research', NAB (ADM/159). Unfortunately, I could not locate Eccles and Jordan's report itself. Therefore, the description of the system is my reconstruction.

17. J. W. Strutt, Baron Rayleigh, *The Theory of Sound*, 2nd edn (New York: Dover 1945), pp. 65–70.

18. First quotation from S. Eccles, 'The Use of the Triode Valve', second quotation from *Annual Report of the National Physical Laboratory for the Year 1919* (Teddington: National Physical Laboratory, 1920), p. 50. This part refers to a work done in 1918.

19. On the request from these bodies see *Annual Report of the National Physical Laboratory for the Year 1919* (Teddington: National Physical Laboratory) for the years 1920 (p. 63), 1923 (p. 84), 1924 (p. 77), 1926 (p. 11), 1928 (p. 13), and D. W. Dye, 'A Self-Contained Standard Harmonic Wave-Meter', *Philosophical Transactions of the Royal*

Society of London. Series A, 224 (1924), pp. 259–301, on p. 300.

20. E.g., S. J. Douglas, *Inventing American Broadcasting, 1899–1922* (Baltimore, MD: Johns Hopkins University Press, 1987), ch. 9, pp. 292–314; C. P. Yeang, 'Characterizing Radio Channels: The Science and Technology of Propagation and Interference, 1900–1935' (PhD dissertation, MIT, 2004), pp. 327–56.

21. E. V. A. [Appleton], 'David William Dye. 1887–1932', *Obituary Notices of Fellows of the Royal Society (1932–1954)*, 1 (1932), pp. 75–8; quotations from Louis Essen, *Time for Reflection*, ch. 2, unpublished memoirs, 1996, in Ray Essen's personal library; L. Hartshorn, 'D. W. Dye, D.Sc., F.R.S.', *Proceedings of the Physical Society*, 44 (1932), pp. 608–10.

22. Dye casually mentioned the use of the tuning fork circuit in the 1919 NPL report (p. 51). Only in the 1921 report did he add that its combination with the multivibrator was 'an improvement first introduced at the Laboratory' (p. 75).

23. NPL Reports for 1922 (p. 83), 1923 (p. 85), H. R. Slotten, *Radio and Television Regulation: Broadcast Technology in the United States, 1920–1960* (Baltimore, MD: Johns Hopkins University Press, 2000), pp. 1–42.

24. The idea of a mechanical tuning fork clock preceded its use for calibration, and it continued to be used also for other ends. D. Pantalony, *Altered Sensations: Rudolph Koenig's Acoustical Workshop in Nineteenth-century Paris* (Dordrecht: Springer, 2009), pp. 100–5.

25. By electronic I mean here circuits whose mechanism relies to some extent on the properties of discrete electrons, as distinct from other electromagnetic properties.

26. D. Dye, 'The Valve-Maintained Tuning-Fork as a Precision Time-Standard', *Proceedings of the Royal Society of London. Series A*, 103 (1923), pp. 240–60, on p. 257; D. W. Dye and L. Essen, 'The Valve Maintained Tuning Fork as a Primary Standard of Frequency', *Proceedings of the Royal Society of London. Series A*, 143 (1934), pp. 285–306, on p. 306.

27. Dye, 'Valve-Maintained Tuning-Fork', pp. 259–60.

28. J. W. Horton, N. H. Ricker and W. A. Marrison, 'Frequency Measurement in Electrical Communication', *Transactions of the AIEE*, 42 (1923), pp. 730–41, quote on p. 731.

29. J. W. Horton, N. H. Ricker and W.A. Marrison, 'Frequency Measurement in Electrical Communication', *Transactions of the AIEE*, 42 (1923), pp. 734–36, Marrison Notebook (NB) 1112-2, pp. 38–41, in AT&T Archives.

30. The tuning fork of AT&T standard 'ran continuously from April, 1923, to May, 1927, except for four intervals totalling about three days'. J. W. Horton and W. A. Marrison, 'Precision Determination of Frequency', *Proceedings of the IRE*, 16 (1928), p. 139.

31. J. W. Horton and W.A. Marrison, 'Precision Determination of Frequency', *Proceedings of the IRE*, 16 (1928), p. 141, Horton, Ricker, and Marrison, 'Frequency Measurements', p. 736; W. A. Marrison, 'The Evolution of the Quartz Crystal Clock', *Bell System Technical Journal*, 27 (1948), pp. 510–88, 528–530, second quote on p. 529; Marrison NB 1444, pp. 97–105 (24.1.25). Bureau of Standards, *Standards Yearbook* (U.S. Govt. Print. Off., 1927), p. 44.

32. S. Katzir, 'Pursuing Frequency Standards and Control: The Invention of Quartz Clock Technologies', *Annals of Science* (forthcoming); C. Stephens and M. Dennis, 'Engineering Time: Inventing the Electronic Wristwatch', *British Journal for the History of Science*, 33 (2000), pp. 477–97.

33. The steam railway seems as a famous example for such a gradual invention.

9 Guralp, Calibrating the Universe: The Beginning and End of the Hubble Wars

1. That this discovery constitutes a milestone is not generally disputed. However, it has recently been claimed that crediting Hubble with this discovery, understood as the discovery of the expanding universe, is not acceptable (H. Kragh and R. Smith, 'Who Discovered the Expanding Universe', *History of Science*, 41 (2003), pp. 141–62). The debate concerning who actually discovered the expanding universe does not have any direct bearing on the issues discussed in this paper. Suffice it to say that, within the period that constitutes the subject of this paper, it was generally accepted by the scientific community that Edwin Hubble was the discoverer of the expanding universe.

2. In this the paper, I use the terms 'experimental' and 'observational' interchangeably. Even though the relationship between these terms is problematic from a general philosophy of science perspective, the context of this paper permits me to pass over this issue.

3. M. L. Humason, N. U. Mayall and A. R. Sandage. 'Redshifts and Magnitudes of Extra-Galactic Nebulae', *Astronomical Journal*, 6:3 (1956), pp. 97–162.

4. A standard candle is a type of astronomical object with an intrinsic brightness that is assumed to be reliably known.

5. The idea of *precision indicators* constituted the heart of Sandage's methodology. As early as 1954, in his Helen Warner Prize Lecture, he argued that searching 'precision indicators' that 'can be isolated and measured' was a key step towards measuring the Hubble constant. See A. Sandage, 'Current Problems in the Extragalactic Distance Scale', *Astrophysical Journal*, 127:3, pp. 513–26 (1954), on p. 514.

6. The most commonly used units for the Hubble constant is kilometre per second per megaparsec, and I follow this convention in this paper. For ease of reading, I will suppress the units when reporting different measured values of the constant for the rest of this paper.

7. H. Chang, *Inventing Temperature: Measurement and Scientific Progress* (Oxford and New York: Oxford University Press, 2004).

8. P. Galison, *Image and Logic: A Material Culture of Microphysics* (Chicago, IL: University of Chicago Press, 1997).

9. The *precision era* in cosmology is usually considered to have begun in the early '90s, when several high precision measurements on the large scale structure of the universe were first made, thanks to crucial technological advances.

10. A. Sandage, 'Cosmology: A Search for Two Numbers', *Physics Today*, 23:2 (1970), pp. 34–41, on p. 34. The two numbers that Sandage is referring to are the *deceleration parameter* and the *Hubble constant*.

11. G. de Vaucouleurs, 'The Extra-Galactic Distance Scale. I. A Review of Distance Indicators: Zero Points and Errors of Primary Indicators', *Astrophysical Journal*, 223 (1978), pp. 351–63.

12. P. Hodge, 'Humason, Mayall, & Sandage's Determination of the Hubble Expansion', *Astrophysical Journal*, 525 (1999), p. 713.

13. A. A. Penzias and R. W. Wilson, 'A Measurement of Excess Antenna Temperature at 4080Mc/s', *Astrophysical Journal*, 142 (1965), p. 419–21.

14. M. L. Humason, N. U. Mayall and A. R. Sandage. 'Redshifts and Magnitudes of Extra-Galactic Nebulae', *Astronomical Journal*, 6:3 (1956), pp. 97–162, on p. 97.

15. As the authors explained in the introductory section, their 'unified treatment of

spectrographic and photometric data follows previous practice by Hubble, who, had he lived, would have participated as the senior author in the analysis and discussion' (Humason et al., 'Redshifts', pp. 97–8).

16. The first section included data from the Mount Wilson-Palomar observatories where Humason worked. The second section by Mayall presented the data he obtained at the Lick Observatory.
17. Humason et al., 'Redshifts', p. 159.
18. Humason et al., 'Redshifts', p. 159.
19. H II regions are composed of low density clouds of partially ionized gas.
20. Humason et al., 'Redshifts', p. 160.
21. Humason et al., 'Redshifts', p. 161.
22. Sandage, 'Cosmology', p. 36.
23. Sandage, 'Cosmology', p. 36.
24. Due to the space limitations of this paper, I cannot go into the question of meaning of calibration in astrophysics research. For a general analytical inquiry on the notion of calibration in science, see L. Soler, F. Wieber, C. Allamel-Raffin, J. Gangloff, C. Dufour and E. Trizio (2013), 'Calibration: A Conceptual Framework Applied to Scientific Practices Which Investigate Natural Phenomena by Means of Standardized Instruments', *Journal for the General Philosophy of Science*, 44 (2013), pp. 263–317.
25. Sandage, 'Cosmology', p. 36. Emphasis added.
26. A. Sandage and G.A. Tammann, 'Steps toward the Hubble Constant. I. Calibration of the Linear Sizes of Extragalactic H II Regions', *Astrophysical Journal*, 190 (1974), pp. 525–38, on p. 525.
27. Sandage and Tammann, 'Steps I', p. 526.
28. Here, P stands for *period*, L for *luminosity* and C for *colour*.
29. G. de Vaucouleurs, 'The Extra-Galactic Distance Scale. I. A Review of Distance Indicators: Zero Points and Errors of Primary Indicators', *Astrophysical Journal*, 223 (1978), pp. 351–63, on p. 351.
30. A. Sandage and G. A. Tammann, 'Steps toward the Hubble Constant. VI. The Hubble Constant Determined from Redshifts and Magnitudes of Remote Sc I Galaxies: The Value of q_0', *Astrophysical Journal*, 197 (1975), pp. 265–80.
31. G. de Vaucouleurs, 'The Cosmic Distance Scale and the Hubble Constant' (Mt. Stromlo & Siding Spring Observatories, Australian National University, Canberra, 1982), pp. 16–17.
32. De Vaucouleurs, 'Cosmic Distance Scale', p. 39.
33. De Vaucouleurs, 'Cosmic Distance Scale', p. 43.
34. This is the reason why the Sandage–de Vaucouleurs debate was referred to as the factor of two controversy.
35. G. de Vaucouleurs, 'Extragalactic Distance Scale, Malmquist Bias and Hubble Constant', *Mon. Not. R. Astr. Soc.*, 202 (1983), pp. 367–78, on p. 368.
36. C. L. Bennett et al., 'Nine-Year Wilkinson Microwave Anisotropy Probe (WMAP) Observations: Final Maps and Results', *Astrophysical Journal Supplement Series*, 208:20 (2013), p. 54.
37. Planck Collaboration: P. A. R. Ade et al., 'Planck 2013 Results. I. Overview of Products and Scientific Results', *Astronomy & Astrophysics*, 571:A1 (2014), 48 pp.
38. W. Freedman et al., 'Final Results from the Hubble Space Telescope Key Project to Measure the Hubble Constant', *Astrophysical Journal*, 553 (2001), pp. 47–72, on p. 50.
39. W. Freedman et al., 'Final Results', p. 50.

40. W. Freedman et al., 'Final Results', p. 50.
41. W. Freedman et al., 'Final Results', p. 51.
42. W. Freedman et al., 'Final Results', p. 51.
43. The HKP obtained the value 72 ± 8 for the constant.
44. See http://hubblesite.org/newscenter/archive/releases/1999/19/text [accessed 23 March 2015].
45. H. Chang, *Inventing Temperature: Measurement and Scientific Progress* (Oxford and new York: Oxford University Press, 2004), p. 224.
46. Chang, *Inventing*, p. 224.
47. Chang, *Inventing*, p. 225.
48. Chang, *Inventing*, pp. 45–6.
49. Chang, *Inventing*, p. 228.
50. P. Galison, *Image and Logic: A Material Culture of Microphysics* (Chicago, IL: University of Chicago Press, 1997), p. 4.
51. Galison, *Image*, p. 4.
52. Galison, *Image*, p. 4.
53. These may include decisions on questions such as which research programs are worth pursuing, which methods are to be followed and how to follow them, in addition to differing approaches to data collection and analysis.
54. Above, I quoted the value from the recent Planck mission. Yet, a team of scientists employing the Hubble Space Telescope to look at the supernova explosions to determine the value of the constant obtained the result of 73.8 ± 2.4 (A. Riess et al., 'A 3% Solution: Determination of the Hubble Constant with the Hubble Space Telescope and Wide Field Camera 3', *Astrophysical Journal*, 730:119 (2011), 18 pp.) Note that the error bars of the two results do not overlap.

10 Hochereau, Measuring Animal Performance: A Sociological Analysis of the Social Construction of Measurement

1. Cf. I. Hacking, *The Emergence of Probability* (Cambridge: Cambridge University Press, 1975); S. Stigler, *The History of Statistics. The Measurement of Uncertainty before 1900* (Cambridge, MA: Harvard University Press, 1986); A. Desrosières, *The Politics of Large Numbers: A History of Statistical Reasoning* (Cambridge, MA: Harvard University Press, 1998).
2. Cf. S. Schaffer, 'Astronomers Mark Time: Discipline and the Personal Equation', *Science in Context*, 2:1 (1988), pp. 115–45; D. Gooding, T. Pinch and S. Schaffer (eds.), *The Uses of Experiment: Studies in the Natural Sciences* (Cambridge: Cambridge University Press, 1989); C. Licoppe, *La Formation de la Pratique Scientifique: Le Discours de l'Expérimentation en France et en Angleterre (1630–1820)* (Paris: La Découverte, 1996).
3. Cf. M. N. Wise (ed.), *The Values of Precision* (Princeton, NJ: Princeton University Press, 1995); T. Quinn and J. Kovalevsky, 'The Development of Modern Metrology and its Role Today', *Phil. Trans. R. Soc.* A, 363 (2005), pp. 2307–27; S. Schaffer, *La Fabrique des sciences modernes* (Paris: Seuil, 2014).
4. A. W. Crosby, *The Measure of Reality: Quantification and Western Society, 1250–1600* (Cambridge, MA: Cambridge University Press 1997).
5. W. Kula, *Measure and Men* (Princeton, NJ: Princeton University Press, 1986).

6. Cf. B. Curtis, 'From the Moral Thermometer to Money: Metrological Reform in Pre-Confederation Canada', *Social Studies of Science* 28/4 (1998), pp. 547–70; G. C. Bowker and S. L. Star, *Sorting Things Out: Classification and its Consequences* (Cambridge: MIT Press, 1999); T. J. Quinn and J. Kovalevsky, 'Measurement and Society, Fundamental Metrology', *C. R. Physique*, 5 (2004), pp. 791–7; J. Yates, *Structuring the Information Age, Life Insurance and Technology in the Twentieth Century* (Baltimore, MD: Johns Hopkins University Press, 2005).

7. K. Alder, 'Making Things the Same: Representation, Tolerance and the End of the Ancien Regime in France', *Social Studies of Science*, 28:4 (1998), pp. 499–545.

8. J. O'Connell, 'Metrology: The Creation of Universality by the Circulation of Particulars', *Social Studies of Science*, 23 (1993), pp. 129–73.

9. A. Mallard, 'Compare, Standardize and Settle Agreement: On Some Usual Metrological Problems', *Social Studies of Science*, 28:4 (1998), pp.571–601.

10. M. Power, 'Counting, Control and Calculation: Reflections on Measuring and Management', *Human Relations*, 57:6 (2004), pp. 765–83; M. Lamont, 'Toward a Comparative Sociology of Valuation and Evaluation', *Annu. Rev. Sociol.*, 38 (2012), pp. 201–21.

11. M. Porter, *Trust in Numbers: The Pursuit of Objectivity in Science and Public Life* (Princeton, NJ: Princeton University Press, 1996); W. N. Espeland and M. Sauder, 'Rankings and Reactivity: How Public Measures Recreate Social Worlds', *American Journal of Sociology*, 113:11 (2003), pp. 1–40.

12. M. Lampland and S. L. Star (eds.),*Standards and their Stories: How Quantifying, Classifying, and Formalizing Practices Shape Everyday Life* (Ithaca, N.Y.: Cornell University Press, 2009).

13. K. R. Scherer, 'What Are Emotions and How Can They Be Measured?', *Social Science Information*, 44:4 (2005), pp. 695–729.

14. L. Dirk, 'The Elements of Science', *Social Studies of Science*, 29:5 (1999), pp. 765–76.

15. T. Moreira, 'Heterogeneity and Coordination of Blood Pressure in Neurosurgery', *Social Studies of Science*, 36 (2006), pp. 69–97.

16. L. Derksen, 'Towards a Sociology of Measurement: The Meaning of Measurement Error in the Case of DNA Profiling', *Social Studies of Science*, 30:6 (2000), pp. 803–45.

17. B. Vissac, *Les vaches de la République. Saisons et raisons d'un chercheur citoyen* (Paris: Quae, 2002).

18. M. Klopcic, R. Reents, J. Philipsson and A. Kuipers (eds), *Breeding for Robustness in Cattle*, EAAP publication 126 (Wageningen: Wageningen Academic Publishers, 2009).

19. Vissac, *Les vaches de la République. Saisons et raisons d'un chercheur citoyen* (Paris: Quae, 2002).

20. J.-L. Mayaud, 'La "belle vache" dans la France des concours agricoles du XIXe siècle', *Cahiers d'histoire*, 42:3–4 (1997), pp. 521–41.

21. This shift from 'traditional' visual and experiential knowledge of livestock animals to the emergence of 'genetic' techniques in livestock breeding is not specific to France but global in Western countries as in the UK. Cf. L. Holloway, C. Morris, B. Gilna and D. Gibbs, 'Choosing and Rejecting Cattle and Sheep: Changing Discourses and Practices of (De) selection in Pedigree Livestock Breeding', *Agric. Hum. Values*, 28 (2011), pp. 533–47.

22. W. N. Espeland and L. S. Mitchell, 'Commensuration as a Social Process', *Annual Review of Sociology*, 24 (1998), pp. 313–43.

23. T. M. Porter, 'Quantification and the Accounting Ideal in Science', *Social Studies of Science*, 22:4 (1992), pp. 633–51.

24. B. Bock and M. M. van Huik, 'Animal Welfare: The Attitudes and Behaviour of European Pig Farmers', *British Food Journal*, 109:11 (2007), pp. 931–44.

25. B. Elzen, F. W. Geels, C. Leeuwis and B. van Mierlo, 'Normative Contestation in transitions "in the making": Animal Welfare Concerns and System Innovation in Pig husbandry', *Research Policy*, 40 (2011), pp. 263–75.

26. B. Theunissen, 'Breeding without Mendelism: Theory and Practice of Dairy Cattle Breeding in the Netherlands 1900–1950', *Journal of the History of Biology*, 41 (2008), pp. 637–76.

27. F. Dagognet, *Réflexions sur la mesure* (La Versanne: Encre Marine, 1993).

28. Dagognet, *Réflexions sur la mesure*.

29. L. Derksen, 'Towards a Sociology of Measurement: The Meaning of Measurement Error in the Case of DNA Profiling', *Social Studies of Science*, 30:6 (2000), pp. 803–45.

30. P. Le Neindre, G. Trillat., J. Sapa, F. Ménissier, J.-N. Bonnet and J.-M. Chupin, 'Individual Differences in Docility in Limousin Cattle', *Journal of Animal Science*, 73 (1995), pp. 2249–53.

31. J. Dewey, *Logic: The Theory of Inquiry* (New York: Holt, Rinehart and Winston, 1938).

32. This concept, originally developed in French Convention Theory, refers to investments which reduce the complexity of things to a set of intermediate elements easier to control and monitor. The idea of investment pays attention to the cost (and the work) of developing such 'forms' (and therefore to the unknown factors and choices of their constitution), which in our case is similar to the successive ways to legitimate the docility test by making it compatible with other animal performance assessments and the various constraints related to its use (cf. L. Thévenot, 'Rules and Implement: Investment in Forms', *Social Science Information*, 23:1 (1984), pp. 1–45).

33. A. Bidet, M. Boutet, T. Le Bianic, O. Minh Fleury, C. Palazzo, G. Rot and F. Vatin, 'Le sens de la MESURE', *Terrains et travaux*, 4 (2003): 'Enquêtes sur l'activité économique' (Paris: ENS Cachan).

11 Crasnow, The Measure of Democracy: Coding in Political Science

1. See C. Boix, *Democracy and Redistribution* (New York: Cambridge University Press, 2003), G. L Munck, *Measuring Democracy: A Bridge Between Scholarship and Politics* (Baltimore, MD: Johns Hopkins University Press, 2009) and J. Gerring, *Social Science Methodology: A Unified Framework* (Cambridge: Cambridge University Press, 2012).

2. H. Chang, *Inventing Temperature: Measurement and Scientific Progress* (Oxford: Oxford University Press, 2004).

3. See B. Russett, *Grasping the Democratic Peace* (Princeton, NJ: Princeton University Press, 1993), for a discussion of the democratic peace hypothesis.

4. J. Gerring, *Social Science Methodology: A Unified Framework* (Cambridge: Cambridge University Press, 2012), p. 156.

5. G. L. Munck, *Measuring Democracy: A Bridge between Scholarship and Politics* (Baltimore, MD: Johns Hopkins University Press, 2009).

6. A. Przeworski, M. E. Alvarez, J. A. Cheibub and F. Limongi, *Democracy and Development: Political Institutions and Well-Being in the World, 1950–1990* (Cambridge: Cambridge University Press, 2000).

7. M. G. Marshall, T. R. Gurr and K. Jaggers. *Polity™ IV Project: Political Regime Characteristics and Transitions, 1800–2009* (Vienna, VA: Center for Systemic Peace, 2010). Dataset at http://www.systemicpeace.org/polityproject.html [accessed 27 March 2015].

8. Munck, *Measuring Democracy*, p. 34.

9. Munck, *Measuring Democracy*, p. 16.
10. P. Paxton, 'Women's Suffrage and the Measurement of Democracy: Problems of Operationalization', *Studies in Comparative International Development*, 43 (2000), pp. 92–111.
11. Paxton, 'Women's Suffrage and the Measurement of Democracy', p. 93.
12. Paxton, 'Women's Suffrage and the Measurement of Democracy', p. 93.
13. R. N. Adcock and D. Collier. 'Measurement Validity: A Shared Standard for Qualitative and Quantitative Research', *American Political Science Review*, 95 (2001), pp. 529–46.
14. N. Cartwright and J. Cat, L. Fleck and T. E. Uebel, *Otto Neurath: Philosophy between Science and Politics* (Cambridge: Cambridge University Press, 1996).
15. C. C. Ragin, *Redesigning Social Inquiry: Fuzzy Sets and Beyond* (Chicago, IL: University of Chicago Press, 2008).
16. Adcock and Collier, 'Measurement Validity', p. 531.
17. A. Kaplan, *The Conduct of Inquiry: Methodology for Behavioral Science* (Scranton, PA: Chandler Publishing, 1964), p. 53.
18. S. Efstathiou, 'How Ordinary Race Concepts Get to Be Usable in Biomedical Science: An Account of Founded Race Concepts', *Philosophy of Science*, 79:5 (December 2012), pp. 701–13.
19. Efstathiou, 'How Ordinary Race Concepts Get to Be Usable in Biomedical Science', p. 707. Emphases in the original.
20. S. Haggard and R. R. Kaufman, 'Inequality and Regime Change: Democratic Transitions and the Stability of Democratic Rule', *American Political Science Review*, 108 (2012), pp. 495–516.
21. J. A. Cheibub, J. Ghandi and J. R. Vreeland, 'Democracy and Dictatorship Revisited', *Public Choice*, 143 (2010), pp. 67–101.
22. M. G. Marshall, T. R. Gurr and K. Jaggers. *Polity™ IV Project: Political Regime Characteristics and Transitions, 1800–2009* (Vienna, VA: Center for Systemic Peace, 2010). Dataset at http://www.systemicpeace.org/polityproject.html [accessed 27 March 2015].
23. G. L. Munck, *Measuring Democracy: A Bridge Between Scholarship and Politics* (Baltimore, MD: Johns Hopkins University Press, 2009), p. 34.
24. J. A. Cheibub, J. Ghandi and J. R. Vreeland, 'Democracy and Dictatorship Revisited', *Public Choice*, 143 (2010), p. 72.
25. S. Huntington, *The Third Wave: Democratization in the Late Twentieth Century* (Norman, OK: University of Oklahoma Press, 1991).
26. C. Boix, *Democracy and Redistribution* (New York: Cambridge University Press, 2003).
27. D. Acemoglu, and J. A. Robinson, 'Democratization or Repression?', *European Economics Review*, 44 (2000), pp. 683–93, 'Theory of Political Transitions', *American Economic Review*, 91(2001), pp. 938–63 and *Economic Origins of Dictatorship and Democracy* (New York: Cambridge University Press, 2006).
28. Haggard and Kaufman, 'Inequality and Regime Change', p. 495.
29. Haggard and Kaufman, 'Inequality and Regime Change', p. 498.
30. H. Chang, *Inventing Temperature: Measurement and Scientific Progress* (Oxford: Oxford University Press, 2004), p. 226.
31. Chang, *Inventing Temperature: Measurement and Scientific Progress*, p. 225.
32. Chang, *Inventing Temperature: Measurement and Scientific Progress*, p. 227.

12 Neswald, Measuring Metabolism

1.   J. A. Harris and F. G. Benedict, 'Biometric Standards for Energy Requirements in Human Nutrition', *Scientific Monthly*, 8 (1919), pp. 385–402, on p. 385.
2.   J. A. Harris and F. G. Benedict, *A Biometric Study of Basal Metabolism in Man* (Washington, DC: Carnegie Institution of Washington, 1919), p. v.
3.   F. L. Holmes, 'The Formation of the Munich School of Metabolism', in W. Colemann and F. L. Holmes (eds), *The Investigative Enterprise. Experimental Physiology in Nineteenth-Century Medicine* (Berkeley, CA: University of California Press, 1988), pp. 179–210; F. L. Holmes, *Claude Bernard and Animal Chemistry: The Emergence of a Scientist* (Cambridge, MA: Harvard University Press, 1974); K. J. Carpenter, *Protein and Energy* (Cambridge, MA: Cambridge University Press, 1994).
4.   G. Rosen, 'Metabolism. The Evolution of a Concept', *Journal of the American Dietetic Association*, 31 (1955), pp. 861–7; E. N. Ackerknecht, 'Metabolism and Respiration from Erasistratus to Lavoisier', *Ciba Symposia*, 6 (1944), pp. 1815–24; F. C. Bing, 'The History of the Term "Metabolism"', *Journal of the History of Medicine and Allied Sciences*, 26 (1971), pp. 158–80.
5.   A. Crawford, *Experiments and Observations on Animal Heat, and the Inflammation of Combustible Bodies. Being an Attempt to resolve these Phenomena into a General Law of Nature* (London: J. Murray, 1779); A. Lavoisier and P. S. de Laplace, 'Mémoire sur la Chaleur', *Mémoires d'Académie des Sciences* (1780), pp. 355–408; P. Dulong, 'De la chaleur animale', *Journal de physiologie expérimentale*, Paris III (1823), pp. 45–52; C. Despretz, 'Recherches expérimental sur les cause de la chaleur animale', *Journal de physiologie expérimentale*, Paris IV (1824), pp. 143–59; P. V. Regnault and J. Reiset, 'Recherches chimique sur la respiration des animaux des diverses classes', *Annales de Chimie et de Physique*, 26 (1849), pp. 299–519; A. Lavoisier and A. Séguin, 'Second Mémoire sur la Respiration', *Annales de Chimie*, 91 (1814), pp. 318–34; E. Mendelssohn, *Heat and Life* (Cambridge, MA: Harvard University Press, 1964).
6.   Protein fulfilled a different function. According to Liebig, muscular work consumed muscle substance, which was then replaced by ingested protein. J. Liebig, *Die Thier-Chemie oder die organische Chemie in ihrer Anwendung auf Physiologie und Pathologie* (Braunschweig: F. Vieweg & Sohn, 1843).
7.   T. Bischoff and C. Voit, *Die Gesetze der Ernährung des Fleischfressers durch neue Untersuchungen festgestellt* (Leipzig: Winter, 1860); C. Voit, *Beiträge zum Kreislauf des Stickstoffs im thierischen Organismus* (Augsburg: J. P. Himmer, 1857).
8.   M. Pettenkofer, 'Ueber die Respiration', *Annalen der Chemie und Pharmacie*, supplement 2, part 1 (1862), pp. 1–52.
9.   M. Pettenkofer and C. Voit, 'Untersuchungen über den Stoffverbrauch des normalen Menschen', *Zeitschrift für Biologie*, 2 (1866), pp. 459–573.
10. Bischoff and Voit, *Gesetze der Ernährung*, p. 38, Pettenkofer and Voit, 'Untersuchungen über den Stoffwechsel'.
11. Described in R. Tigerstedt, 'Respirationsapparate', in R. Tigerstedt (ed.), *Handbuch der physiologischen Methodik*, 3 vols (Leipzig: Hirzel, 1908-14), vol. 1, pp. 71–149, on pp. 84–7; Pettenkofer, 'Ueber die Respiration'.
12. F. L. Holmes, 'The Intake–Output Method in Physiology', *Historical Studies in the Physical Sciences*, 17 (1987), pp. 245–27; F. L. Holmes, 'Carl Voit and the Quantitative Tradition in Biology', in E. Mendelsohn (ed.), *Transformations and Tradition in the Sciences. Essays in Honour of I. Bernard Cohen* (Cambridge: Cambridge University Press, 1984), pp. 455–70.

13. M. Rubner, 'Calorimetrische Methodik', in Medicinische Facultät zu Marburg (ed.), *Zu der fünfzigjährigen Doctor-Jubelfeier des Herrn Carl Ludwig* (Marburg: University Publisher R. Friedrich, 1890), pp. 33–68; M. Rubner, 'Die Kalorimetrie', in R. Tigerstedt (ed.), *Handbuch der physiologischen Methodik*, vol. 1, pp. 150–228, on pp. 201–12.

14. W. O. Atwater and E. B. Rosa, *Description of a New Respiration Calorimeter and Experiments on the Conservation of Energy in the Human Body* (Washington, DC: Government Printing Office, 1899); W. O. Atwater and F. G. Benedict, *A Respiration Calorimeter with Appliances for the Direct Determination of Oxygen* (Washington, DC: Carnegie Institution of Washington, 1905); F. G. Benedict and T. M. Carpenter, *Respiration Calorimeter for Studying the Respiratory Exchange and Energy Transformations of Man* (Washington, DC: Carnegie Institution of Washington, 1910).

15. F. G. Benedict, 'A Comparison of the Direct and Indirect Determination of Oxygen Consumed by Man', *American Journal of Physiology*, 26 (1910), pp. 15–25.

16. This diversity is striking in F. G. Benedict, 'Reports of Visits to Foreign Laboratories', 7 vols, MS, 1906–1932/33, Harvard Medical Library, Francis A. Countway Library of Medicine, Boston, MA, Boxes 6 and 7. Digitised by the Max Planck Institute for the History of Science Virtual Laboratory Project: http://vlp.mpiwg-berlin.mpg.de/library/manuscripts.html#Benedict [accessed 26 January 2014]. See also Tigerstedt, 'Respirationsapparate'.

17. E. Grafe, 'Ein Kopfrespirationsapparat', *Deutsches Archiv für klinische Medizin*, 95 (1909), pp. 529–42; K. Sondén and R. Tigerstedt, 'Untersuchungen über die Respiration und dem Gesammtstoffwechsel des Menschen', *Skandinavisches Archiv für Physiologie*, 6 (1895), pp. 1–224; Tigerstedt, 'Respirationsapparate'; Benedict, 'Reports', intermittently throughout.

18. Schaffer discusses this problem in S. Schaffer, 'Late Victorian Metrology and its Instrumentation: A Manufactory of Ohms', in M. Biagioli (ed.), *Science Studies Reader* (London: Routledge, 1999), pp. 457–78, on p. 457.

19. E. Crawford, T. Shinn and S. Sörlin, 'The Nationalization and Denationalization of the Sciences: An Introductory Essay', in E. Crawford, T. Shinn and S. Sörlin (eds), *Denationalizing Science. The Contexts of International Scientific Practice* (Dordrecht, Boston, MA, and London: Kluwer Academic Publishers, 1993), pp. 1–42, on p. 15.

20. Benedict, 'Reports', especially vols 2 (1910) and 3 (1913).

21. R. Tigerstedt (ed.), *Handbuch der physiologischen Methodik*, 3 vols (Leipzig: Hirzel, 1908–14); E. Abderhalden (ed.), *Handbuch der biologischen Arbeitsmethoden: Gasstoffwechsel und Calorimetrie*, section 10 (Berlin and Wien: Urban & Schwarzenberg, 1926).

22. E. Neswald, 'Strategies of International Community-Building in Early 20th-Century Metabolism Research: The Foreign Laboratory Visits of Francis Gano Benedict', *Historical Studies in the Natural Sciences*, 43 (2013), pp. 1–40.

23. F. G. Benedict, 'Ein Universalrespirationsapparat', *Deutsches Archiv für Klinische Medizin*, 107 (1912), pp. 156–200; F. G. Benedict, 'An Apparatus for studying respiratory exchange', *American Journal of Physiology*, 24 (1909), pp. 345–74.

24. T. M. Carpenter, *A Comparison of Methods for Respiratory Exchange* (Washington, DC: Carnegie Institution of Washington, 1915).

25. H. B. Williams, J. A. Riche and G. Lusk: 'Metabolism of the Dog Following the Ingestion of Meat in Large Quantity', *Journal of Biological Chemistry*, 12 (1912), pp. 349–76.

26. F. G. Benedict and F. Talbot, *Metabolism and Growth from Birth to Puberty* (Washington, DC: Carnegie Institution of Washington, 1921); F. G. Benedict et al., 'The Basal,

Gaseous Metabolism of Normal Men and Women', *Journal of Biological Chemistry*, 18 (1914), pp. 139–55; T. M. Carpenter, *Tables, Factors, and Formulas for Computing Respiratory Exchange and Biological Transformation of Energy* (Washington, DC: Carnegie Institution of Washington, 1921).

27. K. Meeh, 'Oberflächenmessungen des menschlichen Körpers', *Zeitschrift für Biologie*, 15 (1879), pp. 425–57. This was later refined by D. and E. Du Bois, 'A Formula to Estimate the Appropriate Surface Aarea if Height and Weight Be Known', *Archives of Internal Medicine*, 17 (1931), pp. 863–71.

28. M. Rubner, 'Über den Einfluss der Körpergrösse auf Stoff- und Kraftwechsel', *Zeitschrift für Biologie*, 19 (NF, 1883), pp. 535–62; M. Rubner, *Biologische Gesetze* (Marburg: Universitäts-Schriften, 1887). For an earlier statement of this correlation see C. Bergmann, *Ueber die Verhältnisse der Wärmeökonomie der Thiere zu ihrer Grösse* (Göttingen: Karl Bergmann, 1848).

29. Later, Harris and Benedict added the factors of age and gender to the correlation apparatus. J. A. Harris and F. G. Benedict, *A Biometric Study of Basal Metabolism in Man* (Washington, DC: Carnegie Institution of Washington, 1919).

30. G. McLeod, E. Crofts and F. G. Benedict, 'The Basal Metabolism of Some Orientals', *American Journal of Physiology*, 73 (1925), pp. 449–62.

31. See, for example, G. C. Shattuck and F. G. Benedict, 'Further Studies on the Basal Metabolism of Maya Indians in Yucatan', *American Journal of Physiology*, 96 (1931), pp. 518–28; M. Steggerda and F. G. Benedict, 'The Basal Metabolism of Some Browns and Blacks in Jamaica', *American Journal of Physiology*, 85 (1928), pp. 621–33.

32. E. D. Mason and F. G. Benedict, 'The Basal Metabolism of South Indian Women', *Indian Journal of Medical Research*, 19 (1931), pp. 75–98. The entire Report 1 of the *Chinese Journal of Physiology* of 1928 was devoted to this topic.

33. For example, P. Heinbecker, 'Studies on the Metabolism of Eskimos', *Journal of Biological Chemistry*, 80 (1928), pp. 461–75; C. S. Hicks, R. F. Matters and M. L. Mitchell, 'The Standard Metabolism of Australian Aborigines', *Australian Journal of Experimental Biology and Medical Science*, 8 (1931), pp. 69–82; J. Pi-Suñer, 'Basal Metabolism of the Araucanian Mapuches', *American Journal of Physiology*, 105 (1933), pp. 383–8; E. L. Turner and E. Aboushadid, 'The Basal Metabolism and Vital Capacity of Syrian Women', *American Journal of Physiology*, 192 (1930), pp. 189–95.

34. C. Davenport and M. Steggerda, *Race Crossing in Jamaica* (Washington, DC: Carnegie Institution of Washington, 1919).

35. H. Takahira, 'The Basal Metabolism of Normal Japanese Men and Women', in T. Saiki (ed.), *Progress of the Science of Nutrition in Japan* (Geneva: League of Nations, 1926), pp. 11–36.

36. M. Ocampo et al., 'The Basal Metabolism of the Filipinos', *Journal of Nutrition* (1930), pp. 237–44. Most metabolism laboratories in this period were located in temperate regions, with experiments undertaken during the mild seasons.

13 Vera, The Social Construction of Units of Measurement: Institutionalization, Legitimation and Maintenance in Metrology

1. Joint Committee for Guides in Metrology, *International Vocabulary of Metrology* (Sèvres: JCGM, 2008), p. 6.

2. P. Berger and T. Luckmann, *The Social Construction of Reality* (Garden City, NY: Doubleday, 1966), ch. 2.

3. P. Berger and T. Luckmann, *The Social Construction of Reality* (Garden City, NY: Doubleday, 1966), p. 103.
4. Berger and Luckmann, *The Social Construction of Reality*, p. 109.
5. Berger and Luckmann, *The Social Construction of Reality*, p. 67.
6. Berger and Luckmann, *The Social Construction of Reality*, p. 15.
7. On the global process of metrication see H. Vera, *The Social Life of Measures: Metrication in the United States and Mexico, 1789–1994* (PhD Dissertation, New School for Social Research, 2011), pp. 47–120, 494–505.
8. For a panoramic view of these debates see T. Quinn, *From Artefacts to Atoms* (New York: Oxford University Press, 2012), pp. 341–67; R. P. Crease, *World in the Balance* (New York: W.W. Norton, 2011), pp. 249–68.
9. Among the few accounts of the competitors to the metric system are G. Adam, 'Alternatives to the Metric System, Based on British Unit', *Decimal Educator* (1935) 17, pp. 43–6, 18, pp. 11–12, 20–21, 28–29; and R. Zupko, *Revolution in Measurement* (Philadelphia, PA: American Philosophical Society, 1990), pp. 209–25.
10. A. Linklater, *Measuring America* (New York: Walker and Company, 2002), p. 260.
11. M. Maestro, 'Going Metric: How It All Started', *Journal of the History of Ideas*, 3 (1980), pp. 479–86; 'James Watt: Pioneer of Decimal Systems', *Decimal Educator*, 19 (1936), pp. 9–10; *The Papers of Thomas Jefferson*, 41 vols (Princeton, NJ: Princeton University Press, 1971), vol. 16, pp. 619–23.
12. R. Zupko, *Revolution in Measurement*, pp. 124–130; H. A. Klein, *The Science of Measurement* (New York: Dover, 1988), pp. 108–10.
13. On Mendeleev's relevance for Russian metrology, see M. D. Gordin, 'Measure of All the Russians: Metrology and Governance in the Russian Empire', *Kritika*, 4 (2003), pp. 783–815.
14. E. Cox, 'The Metric System: A Quarter-Century of Acceptance (1851–1876)', *Isis*, 13 (1958), p. 77.
15. H. Spencer, 'Against the Metric System', in *Various Fragments* (New York: D. Appleton, 1914), p. 159.
16. On Spencer's anti-metric stance and its legacy see H. Vera, *The Social Life of Measures: Metrication in the United States and Mexico, 1789–1994* (PhD Dissertation, New School for Social Research, 2011), pp. 341–59.
17. See V. F. Lenzen, 'The Contributions of Charles S. Peirce to Metrology', *Proceedings of the American Philosophical Society*, 109 (1965), pp. 29–46. Also: C. S. Peirce, 'Testimony on the Organization of the Coast Survey', in *Writings of Charles S. Peirce*, 8 vols (Bloomington, IN: Indiana University Press, 1982), vol. 3, pp. 149–61.
18. C. S. Peirce, 'Review of Noel's *The Science of Metrology*', in *Writings of Charles S. Peirce, 1886–1890*, vol. 6, pp. 378–9.
19. Peirce, 'Review of Noel's *The Science of Metrology*', vol. 6, p. 378.
20. P. Berger and T. Luckmann, *The Social Construction of Reality* (Garden City, NY: Doubleday, 1966), p. 115.
21. Berger and Luckmann, *The Social Construction of Reality*, p. 71.
22. *Diario Oficial del Salvador*, 27 August 1885.
23. Berger and Luckmann, *The Social Construction of Reality*, p. 128.
24. K. Alder, *The Measure of All Things* (New York: The Free Press, 2002), pp. 326–31.
25. E. Wigglesworth, 'Why Should We Use the Metric System'?, *Western Lancet*, 7 (1878), p. 426.
26. 'The Metric System', *New York Tribune*, 23 December 1902.

27. F. Cardarelli, *Encyclopaedia of Scientific Units, Weights and Measures* (New York: Springer, 2003), pp. 248–9; H. A. Klein, *The Science of Measurement* (New York: Dover, 1988), pp. 82–3.

28. G. F. Kunz, 'New International Metric Diamond Carat of 200 Milligrams', *Science*, 38 (1913), p. 523.

29. Kunz, 'New International Metric Diamond Carat', pp. 523–4.

30. G. F. Kunz, 'The New International Metric Diamond Carat of 200 Milligrams (Adopted July 1, 1913, in the United States)', *Bulletin of the American Institute of Mining and Metallurgical Engineers*, 79–84 (1913), pp. 1225–45.

31. T. Quinn, *From Artefacts to Atoms* (New York: Oxford University Press, 2012), p. 277.

32. Quinn, *From Artefacts to Atoms*, p. 279.

33. Quinn, *From Artefacts to Atoms*, p. 281.

34. F. Cardarelli, *Encyclopaedia of Scientific Units, Weights and Measures* (New York: Springer, 2003), p. 571.

35. J. L. Heilbron, *Weighing Imponderables and Other Quantitative Science around 1800* (Berkeley, CA: University of California Press, 1993), p. 249. On decimalization as a general tendency during the revolution, see K. Alder, *The Measure of All Things* (New York: The Free Press, 2002), pp. 125–59; D. Guedj, *Le mètre du monde* (Paris: Seuil, 2000), ch. 11; Kula, *Measures and Men* (Princeton, NJ: Princeton University Press, 1986), pp. 250–1.

36. The next two paragraphs are based on H. Vera, 'Decimal Time: Misadventures of a Revolutionary Idea, 1793–2008', *KronoScope*, 9:1–2, 2009, pp. 29–48.

37. On the decimalization of time during the French revolution: P. Smith, 'La division décimale du jour', in J.-C. Hocquet and B. Garnier, *Genèse et diffusion du système métrique* (Caen: Editions-diffusion du Lys, 1990), pp. 123–34; M. J. Shaw, *Time and the French Revolution* (PhD dissertation, University of York, 2000), pp. 93–100.

38. 'Decree Establishing the French Era, November 25, 1793 (4 Frimaire, Year II)', in J. H. Stewart, *A Documentary Survey of the French Revolution* (New York: Macmillan, 1951), p. 509.

39. L. Marquet, '24 heures ou 10 heures? Un essai de division décimale du jour (1793–1795)', *L'Astronomie*, 103 (June 1989), p. 287.

40. Vera, 'Decimal Time', pp. 37–46.

14 Ku and Klaessig, A Matter of Size Does Not Matter: Material and Institutional Agencies in Nanotechnology Standardization

1. L. Wittgenstein, *Wittgenstein's Lectures on the Foundations of Mathematics, Cambridge 1939*, ed. C. Diamond (Chicago, IL: Chicago University Press, 1976), p. 105.

2. N. Brunsson, 'Organizations, Markets and Standardization', in N. Brunsson (ed.), *A World of Standards* (Oxford: Oxford University Press, 2002).

3. A review article about nanotechnology regulation and standardization, see Evisa Kica and Diana M. Bowman, 'Regulation by Means of Standardization: Key Legitimacy Issues of Health and Safety Nanotechnology Standards', *Jurimetrics J.* 53 (2012), pp.11–56.

4. One example can be found in National Institute for Occupational Safety and Health 'Filling the Knowledge Gaps for Safe Nanotechnology in the Workplace', *DHHS Publication* No. 2013–101, at http://www.cdc.gov/niosh/docs/2013–101/pdfs/2013–

101.pdf [accessed 7 April 2015]

5. T. M. Porter, 'Objectivity as Standardization: The Rhetoric of Impersonality in Measurement, Statistics, and Cost–Benefit Analysis', in A Megill (ed.), *Rethinking Objectivity* (Durham: Duke University Press, 1992), pp. 197–237; A. Slaton, *Reinforced Concrete and the Modernization of American Building, 1900–1930* (Baltimore, MD: Johns Hopkins University Press, 2001).

6. J. O'Connell, 'Metrology – The Creation of Universality by the Circulation of Particulars', *Social Studies of Science*, 23 (1993), pp. 129–73.

7. T. M. Porter, 'Making Things Quantitative', *Science in Context*, 7 (1994), pp. 389–407.

8. K. M. Olesko, 'The Meaning of Precision: The Exact Sensibility in Early-Nineteeth-Century Germany', in M. N. Wise (Ed.), *The Values of Precision* (Princeton, NJ: Princeton University Press, 1995), pp. 103–34; S. Schaffer, 'Late Victorian Metrology and its Instrumentation: A Manufactory of Ohms', in R. Bud and S. Cozzens (eds), *Invisible Connections: Instruments, Institutions and Science* (Bellingham, WA: SPIE Optical Engineering Press, 1992), pp. 23–56.

9. For the concept of 'center of calculation', see B. Latour, *Science in Action: How to Follow Scientists and Engineers through Society* (Cambridge, MA: Harvard University Press, 1987), pp. 216–57.

10. O'Connell, 'Metrology', p. 137.

11. S. Timmerman and M. Berg, 'Standardization in Action: Achieving Local Universality through Medical Protocols', *Social Studies of Science*, 27 (1997), pp. 273–305.

12. J. Fujimura, *Crafting Science: A Sociohistory of the Quest for the Genetics of Cancer* (Cambridge, MA: Harvard University Press, 1996), pp. 169–70.

13. B. Hazucha, 'International Technical Standards and Essential Patents: From International Harmonization to Competition of Technologies', Society of International Economic Law, Second Biennial Global Conference, University of Barcelona, 8–10 July 2010. Available at SSRN, http://dx.doi.org/10.2139/ssrn.1632567 [accessed 7 April 2015].

14. Concerns about the repeated application of the relativist-constructivist formula and the lack of macro-institutional analysis has been raised within the science studies community. See Z. Baber, 'An Ambiguous Legacy: The Social Construction of the Kuhnian Revolution and its Consequences for the Sociology of Science', *Bulletin of Science, Technology & Society*, 21:1 (2000), pp. 105–19.

15. For example, K. Tamm Hallström, *Organizing International Standardization – ISO and the IASC in Quest of Authority* (Cheltenham: Edward Elgar, 2004)

16. Both authors are nominated experts in the US Technical Advisory Groups for ISO TC229 Nanotechnologies standardization committee and active members in ASTM E56 nanotechnology committee, reviewing standard work items submitted/proposed by US nanotechnology stakeholders or international standard parties.

17. Similar question is raised in S. Yearly, 'Scientific Proofs and International Justice: Why "Universal" Standards of Scientific Evidence Can Undermine Environmental Fairness', in M. E. Rodrigues and H. Machado (eds), *Scientific Proofs and International Justice: The Future for Scientific Standards in Global Environmental Protection and International Trade* (Braga: NES & University of Minho, 2005), pp. 71–87.

18. The empirical data including interview, authorized figures were collected by Ku's field work at the NIST (2005–7), and at the National Cancer Institute (2010–11).

19. L. Daston (ed.), *Biographies of Scientific Objects* (Chicago, IL: University of Chicago Press, 2000)

20. This is the definition by NNI; yet the size range of nanotechnology – whether to use 100 nm as a threshold – has cause significant debates within national and international forum, as the 'size' definition is going to impact practical decision making in nano-product labelling and regulation.

21. R. V. Lapshin, 'Feature-Oriented Scanning Probe Microscopy', in H. S. Nalwa, *Encyclopedia of Nanoscience and Nanotechnology*, 14 (Valencia, CA: American Scientific Publishers, 2011) pp. 105–15.

22. S. Ku, 'Forming Nanobio Expertise: One Organization's Journey on the Road to Translational Nanomedicine', *Nanomedicine and Nanotechnology*, 4 (2012), pp. 366–77.

23. Personal interview with S. Ku, March 2006.

24. Personal interview with S. Ku, April 2006.

25. Personal interview with S. Ku, July 2006.

26. Personal interview with S. Ku, March 2007.

27. S. Ku, 'Room at the Bottom: The Techno-Bureaucratic Space of Gold Nanoparticle Reference Material', in A. Slaton (ed.), *New Materials: Their Social and Cultural Meanings* (Philadelphia, PA: University of Pennsylvania Press, forthcoming).

28. Ku, 'Room at the Bottom'.

29. Personal interview with D. Kaiser, July 2006.

30. Detail information on SRM production and certification can be found at http://www.nist.gov/srm/program_info.cfm [accessed 7 April 2015].

31. For a detail analysis of Weberian bureaucratic model and standardization, see Ku, 'Room at the Bottom'.

32. See the Instruction of Use, in NIST gold RM Report of Investigation, at https://www-s.nist.gov/srmors/view_detail.cfm?srm=8011 [accessed 7 April 2015].

33. Mattli & Büthe, 'Setting International Standards: Technological Rationality or Primacy of Power?', *World Politics*, 56 (2003), pp. 1–42

34. For the US standard policy on government conformity to voluntary consensus standard, see the OMB CircularA-119, at http://www.whitehouse.gov/omb/circulars_a119 [accessed 7 April 2015].

35. A brief history of the ASTEM E56 can be found at http://www.astmnewsroom.org/default.aspx?pageid=849 [accessed 7 April 2015].

36. For the announcement of the ILS, see http://www.astmnewsroom.org/default.aspx?pageid=1122 [accessed 7 April 2015].

37. For more information on ASTM ILS, see http://www.astm.org/ILS/ [accessed 7 April 2015].

38. Personal interview with S. Ku, April 2010.

39. NCL internal result discussion on ASTM ILS201.

40. In biotech and pharmaceutical industry, technicians are required to perform good laboratory practice according to the FDA and OECD guidelines which can be found at http://www.accessdata.fda.gov/scripts/cdrh/cfdocs/cfcfr/CFRSearch.cfm?CFRPart=58.htm [accessed 7 April 2015].

41. For Wittgenstein's rule following thesis and its implication on SSK's meaning finitism, see B. Barnes, D. Bloor and J. Henry, *Sociology of Scientific Knowledge*. (Chicago, IL: University of Chicago Press, 1996); for a full account of the SSK–Wittgenstein rule following application in nanoscale standardization, see S. Ku, 'There's Plenty of Room at the Bottom': A Sociology of Nanodrug Standardization' (PhD dissertation, University of Cambridge, 2009).

42. The NCL assay got the ASTM approval in 2010 to be the ASTM E2524–08 Standard

Test Method for Analysis of Hemolytic Properties of Nanoparticles.

43. 'Putting It to the Test', *ASTM Standardization News*, November–December 2009, at http://www.astm.org/SNEWS/ND_2009/nelson_nd09.html [accessed 7 April 2015].

44. 'Putting It to the Test'.

45. For the information of NKI and whitepaper, see http://www.nano.gov/NSINKI [accessed 7 April 2015].

15 Huber, Measuring by which Standard? How Plurality Challenges the Ideal of Epistemic Singularity

1. J. O'Connell, 'Metrology: The Creation of Universality by the Circulation of Particulars', *Social Studies of Science*, 23 (1993), pp. 129–73, on p. 157.

2. For an introduction to practices of normalization and the very ideal of normality from a statistical point of view, see, for example: T. M. Porter, *The Rise of Statistical Thinking 1820–1900* (Princeton, NJ: Princeton University Press, 1986); J. Link, *Versuch über den Normalismus, Wie Normalität produziert wird*, 3. ergänzte, überarbeitete und neugestaltete Auflage (Göttingen: Vanderhœck & Ruprecht, 2006); L. Huber, 'Norming Normality: On Scientific Fictions and Canonical Visualisations', *Medicine Studies. An International Journal for History, Philosophy and Ethics of Medicine & Allied Sciences*, 3 (2011), pp. 41–52.

3. With 'standards' I refer to a set of norms, that are neither purely technical nor purely social in character: Talking about standards with regard to scientific practices means to acknowledge the very epistemic goals that are associated with practices of standardization. In short, the normative rank of a standard is connected to the question if it allows to realizing an epistemic goal, such as the stabilisation of a phenomena of interest (i.e., chemical purity).

4. N. W. Wise (ed.), *The Values of Precision* (Princeton, NJ, and Chichester: Princeton University Press, 1995).

5. M. Carrier, 'Values and Objectivity in Science: Value-Ladenness, Pluralism and the Epistemic Attitude', *Science and Education*, 22 (2013), pp. 2547–68.

6. Cf. M. Lampland and S. L. Star (eds), *Standards and their Stories: How Quantifying, Classifying, and Formalizing Practices Shape Everyday Life* (Ithaca, NY, and London: Cornell University Press, 2009); J.-P. Gaudillière and I. Löwy (eds), *The Invisible Industrialist, Manufactures and the Production of Scientific Knowledge* (Ipswich: Macmillan, 1998); R. Brownsword and K. Yeung (eds), *Regulating Technologies. Legal Futures, Regulatory Frames and Technological Fixes* (Oxford and Portland, OR: HART Publishing, 2008).

7. Actually, speaking of an ad hoc standard is misleading, considering that the minor epistemic rank of ad hoc standards is inconsistent with the ideal of a genuine scientific norm. The notion 'ad hoc' refers to the specific (local) circumstances under which such norms evolve and, hence, may be regarded as a designated standard. For example, within an experimental setting, the problem arises how to validate the performance of a living organism (i.e., transgenic mouse). Scientists who are confronted with this problem pragmatically tend to ad hoc solutions, especially if there is no standard protocol yet available. Therefore, they introduce a new regime of dealing with the task (= validation of performance). Given that this kind of problem might arise a number of times, scientist may be forced to refer to the pragmatic solution. It is noticeable that

the normative power of ad hoc standards is limited with regard to scientific purposes in scope and reach.

8. In political theory 'recognition' is commonly said to motivate the legitimacy of an authority. For an introduction see: J. Raz, 'Authority and Justification', in J. Raz (ed.), *Authority* (New York: Basil Blackwell, 1990), pp. 115–41; and also H. Arendt, 'What Is Authority'?, in H. Arendt (ed.), *Between Past and Future: Six Exercises in Political Thought* (New York: Viking Press, 1961), pp. 91–141, on p. 91ff: 'Since authority always demands obedience, it is commonly mistaken for some form of power or violence. Yet authority precludes the use of external means of coercion; where force is used, authority itself has failed'.

9. V. G. Hardcastle and C. M. Stewart, 'Localization in the Brain and Other Illusions', in A. Brook and K. Atkins (eds), *Cognition and the Brain: The Philosophy and Neuroscience Movement* (Cambridge and New York: Cambridge University Press, 2005), pp. 27–39.

10. W. Kula, *Measures and Men* (Princeton, NJ: Princeton University Press, 1986), p. 18.

11. Kula illustrates this relation as follows: 'The right to determine measures is an attribute of authority in all advanced societies. It is a prerogative of the ruler to make measures mandatory and to retain the custody of the standards, which are here and there invested with sacral character. The controlling authority, moreover, seeks to unify all measures within its territory and claims the right to punish metrological transgressions'. W. Kula, *Measures and Men* (Princeton, NJ: Princeton University Press, 1986), p. 18.

12. In social epistemology large-scale collaborations and research consortia are critically discussed with regard to data sharing policy, authorship and/or intellectual property.

13. ADNI, a research consortium initiated in 2004 at this stage enclose over fifty research sites in the United States and Canada alone, including 1,000 participants in total. New paradigms within the third research protocol (ADNI 2, 2011–16) are directly related to the overall goal of creating a new generation of reference atlases of the human brain at the Laboratory of Neuro Imaging (LONI) at the University of California, Los Angeles. For an introduction, see: L. Huber, 'LONI & Co: Die epistemische Spezifität digitaler Wissensräume in der kognitiven Neurowissenschaft', *Berichte zur Wissenschaftsgeschichte*, 34 (2011), pp. 174–90. Further information about ADNI is available at http://www.adni.loni.usc.edu [accessed 30 January 2014]. Besides imaging data, research consortia hosted at LONI are providing clinical and biomarker data as much as genome sequencing data. ADNI is said to having obtained whole genome sequences on the largest cohort of individuals related to a single disease. Recently, ADNI and its European partner, the neuGRID project, launched GAAIN, the Global Alzheimer's Association Interactive Network, built on an international database framework with the aim of providing a higher level of global open data sharing using cloud computing. Further information is available at http://www.gaain.org [accessed 30 January 2014].

14. Cf. R. E. Zupko, *Revolution in Measurement: Western European Weights and Measures since the Age of Science* (Philadelphia, PA: The American Philosophical Society, 1990), p. 8ff: 'Occasionally a common, local weight or measure would become so popular that it would gain either wide-spread local acceptance or even unit standardization. A measure of capacity for coal at La Rochelle called the baille, for instance, was eventually considered the equivalent of 1/80 muid. Originally it was any metal or wooden bucket used for carrying water. Most local creations, however, remained unfixed and unregulated ... Thankfully, some local measures never reached either status, but so many did that the disparity between state and local units grew increasingly more troublesome to

the smooth functioning of business and commerce'.

15. H. Chang, *Inventing Temperature: Measurement and Scientific Progress* (Oxford and New York: Oxford University Press, 2004).

16. Most early atlases of the (human) brain stemmed from one individual post-mortem specimen. Atlases take the form of anatomical references, for example the 'Talairach stereotactic system', which still is one of the most popular reference brains in functional imaging. Another quite popular atlas is called 'ICBM152'. It was generated by the Montréal Neurological Institute (MNI template) referring to the average of 152 MRI scans of healthy subjects (normative sample). Lately, the development of so-called probability atlases of brain morphology are strengthened, which are said to allow for a significant higher degree of representation from a statistical point of view.

17. M. Christen, D. A. Vitacco, L. Huber, J. Harboe, S. I. Fabrikant and P. Brugger, 'Colorful Brains: 14 Years of Display Practice in Functional Neuroimaging', *NeuroImage*, 73 (2013), pp. 30–9.

18. P. Keating and A. Cambrosio, 'To Many Numbers: Microarrays in Clinical Cancer Research', *Studies in History and Philosophy of Biological and Biomedical Sciences*, 43 (2012), pp. 37–51.

19. L. Huber, 'LONI & Co: Die epistemische Spezifität digitaler Wissensräume in der kognitiven Neurowissenschaft', *Berichte zur Wissenschaftsgeschichte*, 34 (2011), pp. 174–90.

20. G. C. Bowker, 'Biodiversity Datadiversity', *Social Studies of Science*, 30 (2000), pp. 643–83.

21. Cf. MIAME – further information at htpp://www.mged.org [accessed 30 January 2014].

22. S. Rogers and A. Cambrosio, 'Making a New Technology: The Standardization and Regulation of Microarrays', *Yale Journal of Biology and Medicine*, 80 (2007), pp. 165–78; Keating and Cambrosio, 'Too Many Numbers'.

23. Further information at http: //www.mged.org/Workgroups/MIAME/ miame_2.0.html [accessed 30 January 2014].

24. The 'MIAME protocol' was created by the Microarray Gene Expression Data Society (MGED) – after Rogers and Cambrosio, 'Making a New Technology', p. 165: 'a remarkable bottom-up initiative that brings together different kinds of specialists from academic, commercial, and hybrid settings to produce, maintain, and update microarray standards'. One key purpose of this society is to 'develop standards for biological research data quality, annotation and exchange'. Cf. MIAME at http://www.mged. org [accessed 30 January 2014]. Within this realm, the 'primary purpose of MGED ontology is to provide standard terms for the annotation of microarray experiments'. The MGED Society was founded in 1999. In 2010 the society changed their name in 'Functional Genomics Data Society' (FGED) 'to reflect its current mission which goes beyond microarrays and gene expression to encompass data generated using any functional genomics technology applied to genetic-scale studies of genetic expression, binding, modification (such as DNA methylation), and other related applications'. Cf. FGED at http://www.fged.org [accessed 30 January 2014].

25. J. H. Fujimura, 'Crafting Science: Standardized Packages, Boundary Objects, and "Translation"', in A. Pickering (ed.), *Science as Practice and Culture* (Chicago, IL, and London: University of Chicago Press, 1992), pp. 168–211.

26. S. Leonelli, 'Packaging Small Facts for Re-Use: Databases in Model Organism Biology', in P. Howlett and M. S. Morgan (eds), *How Well Do Facts Travel? The Dissemination of Reliable Knowledge* (Cambridge and New York: Cambridge University Press, 2011),

pp. 325–48; Huber, 'LONI & Co'.

27. For this dichotomic demand of standardization with regard to classification systems, for example the *International Classification of Disease*, see G. C. Bowker and S. L. Star, *Sorting Things Out, Classification and its Consequences* (Cambridge, MA, and London: MIT Press, 1999), especially pp. 139ff. Star and others have suggested that processes of standardization might be illustrated by Bruno Latour's 'immutable mobiles' (cf. B. Latour, *Science in Action, How to Follow Scientists and Engineers through Society* (Cambridge, MA: Harvard University Press, 1987) – see, for example, S. L. Star, 'The Politics of Formal Representation: Wizards, Gurus, and Organizational Complexity', in S. L. Star (ed.), *Ecologies of Knowledge. Work and Politics in Science and Technology* (New York: State University of New York Press, 1995), pp. 88–118, on p. 91: 'These are representations, such as maps, that have the properties of being, in Latour's words, "presentable, readable and combinable" with one another. Such representations also have "optical consistency", that is, visual modularity and standardized interfaces. They are often flattered to make them tractable in combination. They have the important property of conveying information over a distance (displacement) without themselves changing (immutability) ... But no mobiles are completely immutable, as Latour himself has discussed (1987: 241–42). This is because of a "central tension": to be useful, they must be instantiated in, and therefore adapted to, a particular work setting'.

28. Rogers and Cambrosio, 'Making a New Technology'.

29. Rogers and Cambrosio, 'Making a New Technology', p. 174.

INDEX

acceptance, 6, 144, 171, 180, 200, 208

accuracy, 17, 19–21, 68–9, 84, 91–2, 111, 113–14, 118–22, 126, 135, 138, 149, 153, 155–6, 158, 162, 165, 169–70, 194, 207

acknowledgement, 208–9

actor
non-human, 8, 190, 205–6

adequacy, 83, 102–4, 109, 156, 159–60

agreement, 8, 13, 15, 19, 32–5, 38, 66, 71, 82, 110, 112, 117, 143, 155, 185, 187, 190, 200, 207

algebra, 61–3

Alzheimer's Disease Neuroimaging Initiative, 211

American Bureau of Standards (BoS), 121

American Society for Testing and Materials International (ASTM International), 192, 200–1, 203–4

American Telephone and Telegraph Company (AT&T), 111

analysis
air analysis, 164
chemical analysis, 166
dimensional analysis, 59
multidimensional (MDA), 39–51
of datasets, 135, 138, 202
of error, 83
respiration gas analysis, 164
sociological analysis, 176
statistical, 2, 83, 150, 195
uncertainty analysis, 83

apparatus, 119, 161–7, 170–2, 181–3
respiration, 163, 168, 170

application, 3, 8, 23, 28–9, 39, 42, 47, 62, 64, 69, 71, 76, 83, 101, 118, 153, 162, 167, 171, 192–3, 200, 210–11
military, 117
approximation, 47, 154–5

archaeology, 7, 39–41, 44, 47–8, 50, 137

archaeometry, 51

artifact, 29, 31, 36, 44, 47, 106, 109, 140, 178, 187

astronomy, 58, 128

authority, 6, 8, 24, 134, 144–5, 179, 181, 190–2, 197, 200, 203, 207, 209–11, 213, 215

average, 157, 170

Berger, Peter L., 8, 173–5

biochemistry, 71

biology
molecular, 212
synthetic, 25–37
bio-industry, 190, 209

Bridgman, P. W., 25, 38

British National Physical Laboratory, 111

calibration, 95–102, 105, 107
distance, 136
test, 102–4, 107–10
theory of, 7

calorimeter, 164–5

Cambrosio, Alberto, 40, 212

Cartwright, Nancy, 153

certainty, 13–14

Chang, Hasok, 126, 136–7, 150, 158–60

chemistry, 41, 71, 163
physiological, 8, 163, 165, 168

classification, 3, 13, 21, 39, 90, 150, 209

clock
    quartz, 111, 122–3
    tuning fork, 111, 120–3
code, 8, 42, 146–7, 149–53, 155–60, 212
coherence, 26, 175
Collins, Harry, 231
colour, 13, 16, 21, 212
commensurability, 171
communication, 5, 55, 112, 116, 118, 120, 122, 124, 183
    telecommunication, 112, 117, 120, 124
    wireless, 112
community, 21, 25, 37, 134
    building, 6–7, 206
    scientific, 33, 138, 208–10
    speech, 44
complexity, 40, 162–3, 165
concepts
    conceptualization, 97, 151–3, 155, 205
    latent concepts, 8, 149, 151–2, 154
    metrological concepts, 25, 38
consensus, 32–5, 38, 57, 70, 191, 200–1, 204, 214
consistency, 32, 71, 152, 202
    metrological, 32–3
constant
    fundamental constants (of nature), 65, 187
    Hubble constant, 8, 125–31, 133–8
control, 3, 27, 40, 44, 142, 146, 163–5, 170, 174, 191–2, 194, 200–1, 203–4, 214
convention, 173
coordination, 7, 11, 13, 15, 22–3, 55, 57–8, 124, 179, 195–6, 203
cosmology, 125–7, 136–8
credibility, 3, 191, 196–7, 201
culture
    material, 8, 126, 137–8

data
    database, 30, 135, 211
    data-driven research, 214
    dataset, 42, 45–6, 129, 149, 151–3, 156–9
    mining, 2
    model, 83, 85
    qualitative, 26, 56–7, 60, 142–3, 145–6, 149, 157

quantitative, 1, 30–1, 35–7, 54, 56, 60, 78, 88, 93, 141–3, 146, 149, 194
decision making, 70, 79, 191, 205
definition, 71, 76, 80, 82, 85, 87–9, 91, 101–2, 107, 130, 144, 146, 153–4, 157, 182–3, 185, 194, 214
    atomic, 186
democracy, 8, 149–60
dependency, 63, 70, 77, 103, 145, 152–3, 193
determination, 13, 75, 82, 93, 101, 125, 127–8, 130, 136, 169
dimension, 53
    dimensional analysis, 59
discipline, 4, 28, 35, 39–40, 44, 50, 70, 191, 197
disciplinary boundaries, 1, 4, 6, 44
drugs
    nanodrugs, 202

engineering, 26–7, 29, 33, 35, 37, 119
    biological, 6, 33
    electrical, 117
    genetic, 27–9, 31
    US Synthetic Biology Engineering Research Centre, 26
epistemology, 2, 4, 7, 25, 34, 40, 57, 138
equations, 5, 53–6, 58–67, 85, 89, 91, 93
error, 2
    measurement, 82–3
    theory, 81
ethnography, 26, 40, 44
evaluation, 8, 30, 74, 98, 145–6, 155
evidence, 191
exactness, 114, 183
exchange, 68, 145, 164, 171, 190
experiment, 166
    experimental design, 213
    experimental reduction, 162
    experimentation, 2, 165
    Minimum Information About a Microarray Experiment (MIAME), 213–14
expert, 69, 116, 123, 209
expertise, 3, 6, 113, 118–19, 124, 141–2, 144–5, 161, 183, 193, 206, 210

formalization, 3, 7, 39, 73, 187
Fourier, Charles, 54–5, 58, 60, 224
Fraassen, Bas van, 11, 22–4

frequency, 7, 111–24
  input, 113
Fujimura, Joan, 3, 190, 214

Galison, Peter, 135, 137
Gauss, Carl Friedrich, 81
General Conference on Weights and Measures (CGPM), 62
Guide to the Expression of Uncertainty in Measurement (GUM), 81, 87, 92

Haggard, Stephan, 155–60
harmonization, 71, 81, 206
history, 2, 44, 70, 95, 125–7, 161, 174, 178–9, 185
Hubble, Edwin, 125
Huntington, Samuel, 153, 156

idealization, 21, 28, 35, 46, 70, 75, 81–3, 89, 91–2, 96, 142, 208, 210, 214
identification, 26, 89–92, 97, 128–9, 147, 149, 158
implementation, 39, 64, 141, 183, 197, 208–9, 213
infrastructure, 139, 170, 176, 196–7, 203, 205–6
institution, 14, 174–5, 192, 201, 203
institutionalization, 8, 140–1, 174–5, 187
International Committee for Weights and Measures (CIPM), 185
International Electrotechnical Commission (IEC), 70–1
International Federation of Clinical Chemistry and Laboratory Medicine (IFCC), 71
International Laboratory Accreditation Cooperation (ILAC), 71
International Organization for Standardization (ISO), 70–1
International Organization of Legal Metrology (OIML), 70
International Union of Pure and Applied Chemistry (IUPAC), 71
International Union of Pure and Applied Physics (IUPAP), 71
International Vocabulary of Metrology (VIM), 7, 70–82, 87, 90, 92, 173
iteration

epistemic, 8, 138, 150, 158–9

Joint Committee for Guides in Metrology (JCGM), 72
justification, 4, 21, 23, 61, 69, 136, 205

Kant, Immanuel, 1
Keating, Peter, 40, 212
knowledge
  knowability, 88, 92
  production, 137, 142, 149–50, 158, 190, 203
Kula, Witold, 140, 210

laboratory
  Carnegie Nutrition Laboratory, 161, 166, 168
  industrial research, 112
  laboratory medicine, 71
language, 175
  artificial, 47
Laplace, Georges, 41, 43, 81
Latour, Bruno, 190
law
  law of nature, 5
  phenomenological, 90, 92
legitimacy, 41, 144, 146–7, 187
Liebig, Justus, 163
Luckmann, Thomas, 8, 173–4
luminometer, 31

mathematics, 1, 7, 12–13, 41, 54, 56, 61, 68, 139
mathematization, 54, 60
Maxwell, James Clerk, 53, 60–4
measurand, 74–80, 82–3, 87–91
measurement
  axiomatic theory of, 69
  criteria, 146
  frequency, 113
  range, 130
  science, 7, 69–72, 80–1
  theory, 2, 23, 86
  unit of, 5, 12, 14, 23, 173, 176–7, 184
medicine
  biomedicine, 212, 214–15
  nanomedicine, 193, 196
metabolism, 161–72
metric, 8, 149

biometric, 162
equivalency, 185
metrication, 179, 181–4
reform, 55, 58, 177
system, 176–87
metrology, 2, 5–7, 11, 15, 22–5, 30–2, 34–7, 40, 44, 70–2, 81–6, 92, 116, 172, 176, 178–9, 184, 186, 190, 194
micrometer, 87
microscope, 108–9, 192
model, 14, 17, 50–1, 63–4, 76–7, 80, 84–5, 89–90, 99–101, 104–9, 127, 133, 136, 160, 166, 214
modelling, 2, 30
multivibrator, 112–16, 119, 123

nomenclature, 178, 213
norms, 6, 8, 28, 34, 37, 161–2, 167–71, 184, 207–8, 210–12, 214
normalization, 3, 25, 207
normative, 4, 6, 28, 139–40, 174, 207–8, 210
nutrition, 8, 161, 163–4, 167–8

object, 89, 91, 106, 129, 203
conceptual, 98
epistemic, 7
material, 197
of calibration, 97
of measurement, 74
scientific, 194
standard object, 192–3, 195–7, 200
objectivity, 46, 67, 79, 140, 144–6, 174–5, 189–90, 207
observation, 5, 56, 74, 89, 129–30, 153
organization, 32, 55, 150, 176, 183, 187, 192, 200, 203, 214

parameter, 84, 128–9, 193, 197
participation, 151–2, 191–2, 211
Peirce, Charles S., 180
phenomenology, 91
phenomenon, 25, 73, 77, 83, 155, 171, 179, 210, 212
natural, 124
physical, 122
philosophy, 2–3, 11, 15, 86, 95

physics, 2, 5, 50, 53–64, 66–8, 71, 90, 93, 112, 117, 121, 135, 193
physiology, 8, 163, 166–8
pluralism, 149–50, 155, 200, 203
plurality, 212, 214
political science, 149–50, 155, 158, 191
population, 145, 157, 170, 176–7, 182, 187, 204
animal, 142
power, 9, 123, 137, 140, 142, 159–60, 175, 180, 187, 190, 201, 204, 207, 210, 214
explanatory, 159–60
precision, 2, 7, 69, 98, 101, 114–15, 118–20, 124, 126, 131, 135, 137–8, 165–6, 172, 185, 189, 192, 194, 197, 201–5, 207, 210
prehistory, 41
probability, 79, 83–5, 92
property, 5, 56, 61, 66, 72–4, 77–8, 89, 107–9, 154–5, 193, 202, 204, 233
protocol, 83, 145, 194, 201–4, 208–9, 212–13
prototype, 97, 105–8, 161, 178
purity, 114, 184

quantity
calculus, 61–3
physical, 81
quantification, 8, 30, 139–40, 142, 149
value, 75–6, 78, 80
rationalization, 50

realism, 75, 86, 93
scientific, 81
reference data, 209, 212, 214
reference material, 195
Gold Nanoparticle Reference Material (gold RM), 192, 196–8, 200–5
regulation, 2–3, 118, 189, 191, 200
representation, 2, 6–7, 23, 54, 84, 106, 142, 149, 153, 207
mathematical, 55
reproduction, 5, 175
research, 125, 131, 154–8, 171, 179, 209
acoustical, 117
breeding, 144
mechanical, 111
military, 112

policymaking, 206
programme, 132, 169–70
radio, 113
robustness, 140, 142, 197–8, 204
rule
   of grammar, 12
   rule following, 6, 11, 14–16, 23, 191, 198, 203, 205
   rule following , 189

Sandage, Allan, 126–7, 131–2, 134–5, 137–8
scale, 56, 59
   atomic, 193
   Cosmic Distance Scale, 133
   nanoscale, 192–3
scheme, 125, 128, 132, 185
   Cartesian, 62
Science and Technology Studies (STS), 191–2, 206
scope, 71, 80, 162
significance, 3
skill, 161, 165
social construction, 139, 147, 173–5, 177, 190
social science, 52, 140, 150–1, 155, 159
sociology, 4, 24–5, 34, 44, 50, 176–7, 191–2
sovereignty, 8, 181
Spencer, Herbert, 180
stability, 119, 121
stabilization, 2–3, 97, 105–6, 109, 162, 207–8, 211, 214
standard
   apparatus, 162
   candle, 125, 128, 131
   certified, 195, 201
   collective, 204
   deviation, 3, 201
   experimental subject, 167
   format, 7, 208–11
   gauge, 207, 210
   mass, 98
   metastandards, 208, 215
   methods, 190
   metre, 12
   metrological, 31
   official, 178
   setting, 209

singularity of, 210
testing material, 201
users, 204
standardization, 78, 149, 161, 171, 194, 201, 203, 205–7
   international, 167
   of parts, 30
   of procedure, 172
   ontology of, 203
   process, 204
   standardized parts, 29
   technical, 206
statistics, 39, 41–2, 46, 48, 139, 143
   Bayesian statistics, 83–5
studies, 170
   Inter-Laboratory Studies (ILS), 192, 201–5
   multicentre studies, 211–13
system
   Bell system, 120–3, 231
   classification, 250
   classificatory, 36
   concept system, 70, 73, 80
   experimental, 196
   labelling, 213
   metric, 177, 186
   metrological, 78
   of knowledge, 136, 159
   of measurement, 35
   of representation, 58, 66
   of units, 65, 86
   of weights and measures, 176
   semiotic, 45
   sign system, 175
   typological, 43
systematization, 154

technology, 2–3, 28, 32, 120, 209
   biotechnology, 28, 31, 190
   committee, 200
   electronic, 111–12
   innovation, 189
   manufacturing, 3
   mass, 124
   nanotechnology, 8, 190, 192–3, 196, 200, 202–3, 205–6
   regulation, 189
   stakeholders, 190

technoscience, 192
  wireless, 123
technoscientific system, 189, 205
telegraphy, 112, 117
  radio, 112–13, 116–18
  telescope, 135
temperature, 56, 76, 87, 91, 98, 119, 150, 154–5, 158–9, 164–5, 168–9, 210, 212
thermometer, 109
time, 21, 53, 59–60, 67, 115–22, 124, 165–6, 176, 184–7, 207
traceability, 78, 182
trust, 3, 33, 140, 203, 205
truth, 20, 85–6, 92–3, 155
tuning fork, 121. *See also* clock: tuning fork, 112, 114–17, 119–20
type, 98, 156–7
typology, 43, 45, 51
  uncertainty, 67, 71–2, 75, 78–80, 82–3, 93, 98–9, 101, 135, 193, 195, 203
  definitional, 87–92
  measurement, 91

uniform, 6, 25, 29–30, 33, 36–7, 97, 207, 211, 214
  parts, 29
  regime, 141
uniformity, 141, 149, 171, 179, 189–90
uniqueness, 86–9, 91, 106, 128, 130, 162, 166, 201
units, 58–9, 182, 184
  base, 63, 65, 177–8
  electromagnetic, 116

International System of Units (SI), 65, 67, 95–6, 139, 178, 187
  social construction of units. *See* social construction
universality, 15, 19, 57, 97, 103, 142–3, 147, 151–3, 162, 170–1, 180, 189–91, 211
universe, 127, 137
US Food and Drug Administration (FDA), 193–4, 196, 204, 208–9
US National Institute of Standards and Technology (NIST), 192–8, 200–1, 203–4, 206

validation, 3, 37, 192, 207–8, 210
value
  numerical value, 85, 98–9, 128
  scientific value, 124
  true value, 75, 81–3, 85–9, 92, 196, 201
variable, 63, 96, 130, 168
Vaucouleurs, Gérard de, 125–6, 131–5, 137–8
verification, 58, 182, 208
virtue, 61, 86, 155, 159
  epistemic virtues, 155, 159
vocabulary, 6–7, 19, 70–1

Wittgenstein, Ludwig, 6, 11–20, 23–4